DIAGNOSIS FOR CLASSROOM SUCCESS

Making Anatomy + Physiology Come Alive

Teacher Edition

Nicole H. Maller

DIAGNOSIS FOR CLASSROOM SUCCESS

Making Anatomy + Physiology Come Alive

Teacher Edition

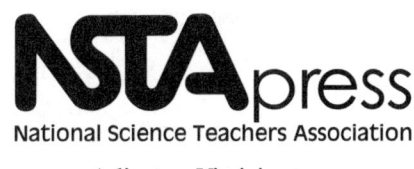

NSTApress
National Science Teachers Association

Arlington, Virginia

National Science Teachers Association

Claire Reinburg, Director
Jennifer Horak, Managing Editor
Andrew Cooke, Senior Editor
Wendy Rubin, Associate Editor
Amy America, Book Acquisitions Coordinator

ART AND DESIGN
Will Thomas Jr., Director
Joe Butera, Senior Graphic Designer, cover and interior design
Images courtesy of ThinkStock.

PRINTING AND PRODUCTION
Catherine Lorrain, Director

NATIONAL SCIENCE TEACHERS ASSOCIATION
David Evans, Executive Director
David Beacom, Publisher
1840 Wilson Blvd., Arlington, VA 22201
www.nsta.org/store

Copyright © 2013 by the National Science Teachers Association.
All rights reserved. Printed in the United States of America.
16 15 14 13   4 3 2 1

NSTA is committed to publishing material that promotes the best in inquiry-based science education. However, conditions of actual use may vary, and the safety procedures and practices described in this book are intended to serve only as a guide. Additional precautionary measures may be required. NSTA and the authors do not warrant or represent that the procedures and practices in this book meet any safety code or standard of federal, state, or local regulations. NSTA and the authors disclaim any liability for personal injury or damage to property arising out of or relating to the use of this book, including any of the recommendations, instructions, or materials contained therein.

PERMISSIONS
Book purchasers may photocopy, print, or e-mail up to five copies of an NSTA book chapter for personal use only; this does not include display or promotional use. Elementary, middle, and high school teachers may reproduce forms, sample documents, and single NSTA book chapters needed for classroom or noncommercial, professional-development use only. E-book buyers may download files to multiple personal devices but are prohibited from posting the files to third-party servers or websites, or from passing files to non-buyers. For additional permission to photocopy or use material electronically from this NSTA Press book, please contact the Copyright Clearance Center (CCC) (www.copyright.com; 978-750-8400). Please access www.nsta.org/permissions for further information about NSTA's rights and permissions policies.

Library of Congress Cataloging-in-Publication Data
Maller, Nicole H.
  Diagnosis for classroom success: making anatomy and physiology come alive / by Nicole H. Maller. — Teacher edition.
      pages cm
  Includes index.
  ISBN 978-1-936959-52-5
  1. Diagnosis—Study and teaching. 2. Human anatomy—Study and teaching. 3. Human physiology—Study and teaching. 4. Diagnosis. I. Title.
  RC71.3.M273 2013
  612.0076—dc23
                                  2013009441

CONTENTS

About the Author vii

Acknowledgments ix

1 BEFORE

Chapter 1: Overview 3

Chapter 2: Aligning to Standards 9

Chapter 3: Teacher Prep Made Easy 19

2 DURING

Chapter 4: Teacher Guide for
Earning Your White Coats: Medical School Research 31

Chapter 5: Teacher Guide for
What's Wrong With Me, Doc? Analyzing Medical Records 43

Chapter 6: Teacher Guide for
Let's Diagnose Them, Lab 1: Urinalysis 59

Chapter 7: Teacher Guide for
Let's Diagnose Them, Lab 2: Digestive By-Products and Body Mass Index Analysis 67

Chapter 8: Teacher Guide for
Let's Diagnose Them, Lab 3: Blood Smears 79

Chapter 9: Teacher Guide for
Let's Diagnose Them, Lab 4: HIV Test 87

CONTENTS

2 DURING

Chapter 10: Teacher Guide for
Let's Diagnose Them, Lab 5: Lung Capacity 97

Chapter 11: Teacher Guide for
Let's Diagnose Them, Lab 6: Hormone Test 107

Chapter 12: Teacher Guide for
Emergency! Lab 7: Performing Surgery 115

Chapter 13: Teacher Guide for
The Ominous Phone Call and Evaluating the Docs 125

3 AFTER

Teacher Survey 141

Index 143

Note: **The complete text of the *Student Edition* follows the Index.**

About the Author

Nicole H. Maller received a B.S. in Teaching Biology 7–12 from New York University in 2006 and M.A. in Science Education from New York University in 2010. Her career in education began in Williamsburg, Brooklyn at The Green School: An Academy for Environmental Careers. A year later, she relocated to Manhattan and worked at Vanguard High School, where she continues to teach Living Environment to 10th graders and a Biopsychology course she developed specifically for 11th and 12th graders. During her summers, Nicole teaches Introductory Chemistry and Introductory Forensics at Columbia University's six-week Upward Bound program to first-generation college-bound students. She also tutors middle school and high school students in Manhattan.

Acknowledgments

Vanguard High School
for providing teachers the freedom to teach students the best way they know how

Catherine Bell
for helping me make this vision come to life in the classroom

NYU Professors
Dr. Pamela Fraser-Abder, Catherine Milne, Jason Blonstein, and Bob Wallace
for your guidance and professional insight

Tal Savariego
for your continuous support and editing skills

Jaimie Glick, M.D.
for evaluating my 'Docs' at round tables

Adam Handler, M.D.
for providing feedback and editing for medical accuracy

Family and Friends
for listening to and believing in my ideas

1 BEFORE

1
2
3

STUDENT EDITION

Chapter 1
Overview

The Approach to This Book

Narratives

Everyone loves a good story, which is why narratives are so essential to our curricula design. "Stories revolve around what matters to people. They are human-centered in their essence and we are in consequence generally drawn to them" (Denning 2009). In fact, a great narrative consists of characters we can identify with, a sequence of events we can follow, and a problem that has potential to reach a resolution (Gilbert, Hopkins, and Cooper 2005). Considering the growing diversity of classrooms today, narratives—found across all cultures—are commonly used to stimulate learning (Barthes 1978; Milne 2011).

Narratives also have a tendency to connect to our everyday experiences. When narratives are well written, readers are able to relate to the characters' goals, obstacles, or their actions to overcome those obstacles, thus, eliciting an emotional response in the readers (Graesser, Singer, and Trabasso 1994; Milne 2011). Milne states that "because human knowledge about these actions, goals, events, and emotions is deeply embedded in our ways of being, we are likely to be more responsive to reading from a narrative text than expository one, such as the decontextualized text found in a textbook" (2011, p. 956).

Many educators mistakenly presume that narratives diminish the level of complexity and analytical thinking expected in a middle school- or high school–level science course. Conversely, narratives, when used correctly in a classroom setting, should not replace abstract ideas, but rather assist students in conceptualizing them (Denning 2009). In fact, researchers who promote the use of storytelling argue that they "supplement [analytical thinking] by enabling [students] to imagine new perspectives and new worlds, and [are] ideally suited to communicating change and stimulating innovation" (Denning 2009). Furthermore, Bruner (2004) argues, "the ways of telling and the ways of conceptualizing, [have] become so habitual that they finally become recipes for structuring experience itself, [and] for laying down the routes into memory" (p. 708). Considering this habit of human nature, students who, for a variety of reasons, have difficulty making sense of textbook- and lecture-based jargon may benefit most from narratives. Evidently, stories do, in fact, help students retain information for greater periods of time and better understand concepts (Kreps-Frisch and Saunders 2008).

Role-Play

Role-play can be an invaluable tool for teaching science concepts. What purpose does knowledge serve if our students are unable to apply it? Although role-play comes in a variety of forms, ranging from gaming to free-play, the labs in *Diagnosis for Classroom Success* tap into the simulated experience (McSharry and Jones 2000). Students complete an array of hands-on activities, allowing them to take on the identity of working professionals. In doing so, Craciun (2010) argues, students are not only exposed to potential science careers, but are able to refine their communication, collaboration, and leadership skills, all of which are skills emphasized within the Common Core State Standards.[1]

Although race in education is often perceived as a sensitive topic, Ladson-Billings and Tate (1995) contend that minority groups, in particular, have a tendency to "internalize the stereotypic images that certain elements of society [perhaps, academic institutions] have constructed in order to maintain their power" (p. 57). Regardless of which marginalized group, whether based on race, gender, ethnicity, economic class, or sexual orientation, role-play aids in the diminishment of these stereotypes and helps to foster new self-identities. In fact, role-play provides an opportunity for those who may have otherwise been pigeonholed into stereotypical vocations to envision themselves in alternative positions of society and empowers them to break away from traditional or expected roles in the pursuit of science.

While role-play exhibits several classroom benefits, it is equally important for educators to acknowledge how classroom simulations differ from the real techniques used by professionals. Before beginning any of our workbooks, educators should clarify that these activities and labs are merely imitations. In this book, those imitations range from the body fluids collected from patients to the diagnostic procedures performed. For teachers who may be interested in showing their students a side-by-side comparison of the medical techniques used within this book, we have provided a chart in Chapter 3 (pp. 25–27).

---

1. For more information on how this book relates to the Common Core State Standards, see Chapter 2.

Pop Culture and Cultural Relevancy

Pop culture undoubtedly infiltrates the lives of youth today. On average, students spend a total of seven hours a day (yes, you heard us correctly!) in front of electronic devices, including computer monitors, phones, television screens, or headphones (Duncan-Andrade 2004). Since students embrace these varying modalities, we, too, have embraced them into our pedagogical approach. In fact, Howard (2003) explains that "culturally relevant pedagogy recognizes the connection between culture and learning, and sees students' cultural capital as an asset and not a detriment to their school success" (p. 198). Milne (2011) argues that a child's everyday language is another aspect of cultural identity that science educators can tap into when cultivating scientific thinking. According to Varelas and colleagues (2008):

> [W]e believe that this is not an "either-or" issue. Learners need to be introduced to academic Discourse (language, ideas, ways of being, acting, and thinking) (Gee 1991), but they need to use it in ways that allow them to bridge this Discourse with their life world Discourse and everyday experience if they are to "own" science. (p. 67) (quoted in Milne 2011, p. 959)

Whether the emphasis is on technology itself, pop culture, or the life experiences of students, we constantly strive to build on students' strengths to promote classroom participation through familiarity and confidence building.

Similarly, Kirkland (2003), when explaining how teachers can improve their pedagogical approach, states that teachers need to "understand the experiences and perspectives [students] bring to educational settings and be responsive to the cultures of different groups in designing curriculum, learning activities, classroom climate, instructional materials and techniques, and assessment procedures" (p. 134). Our activities, essentially, do just that, and seek to facilitate this paradigm shift for teachers hesitant in making such a drastic pedagogical transition in their own line of work.

We are dedicated to the development of curricula that drives student interest through cultural relevancy. Therefore, we have made it our goal to include multiple cultural references to motivate and engage your students.

Why This Approach?

We realize all students' learning and life goals are different. We understand some students within the science classroom are optimistic about pursuing a career in science while others cannot comprehend the value in learning science. Our curricula promote the four needs in modern science education as identified by The National Science Foundation's Project Synthesis in 1978: personal needs, societal needs, career awareness, and academic preparation (Arieli 2007). Rather than focus science education solely on the select few pursuing higher education and careers in science, we foster individual awareness of health and environmental issues at personal and societal levels.

How Do We Ensure the Curriculum Covers State-Mandated Material?

When writing this book, we kept the following four questions in mind:

- What concepts do students need to understand?
- What story can we create to link these concepts together?
- How can we bring pop culture and cultural relevancy into this story?
- What activities and assessments can be used to demonstrate understanding?

Our curriculum is aligned with the Framework[2] used to develop the Next Generation Science Standards (NGSS) and The Core Curriculum State Standards (CCSS) so that they can be easily adapted into any science classroom in the United States. Prior to writing, we identified the science concepts covered within each unit. Once the concepts have been determined, an engaging plot (with pop culture references and cultural relevancy) is created that links the concepts together through a narrative. Within that narrative, students take on a character and engage in hands-on work either to simulate a career or to help demonstrate their understanding. Students, therefore, have the ability to show what they know through multiple methods including laboratory activities, visual projects, oral presentations, written assignments, and so forth.

How Does This Approach Fit Into My Classroom Routines?

We are fully aware the school day is structured differently for science teachers across districts. Some teachers only have 45 minutes of face-to-face time with their students, while others may have 90-minute blocks. These activities were designed to supplement and simplify the work teachers currently do in their classrooms. In fact, each chapter of this book contains "Before the Lesson," "During the Lesson," and "After the Lesson" recommendations to help guide instruction, a complete list of supplies,1 as well as a pacing chart to ease teacher preparation time.

---

2. At the time of writing, the Next Generation Science Standards are still being drafted. In the meantime, our curriculum is aligned to National Academies Press' *A Framework for K–12 Science Education: Practices, Crosscutting Concepts, and Core Ideas* (2012).

What Is the Best Way to Use This Teacher Edition?

Before jumping into the labs, we suggest you review Chapters 2 and 3. Chapter 2 lays out how the student workbook is aligned to the NGSS and CCSS, and Chapter 3 provides you with a list of supplies required for implementing all activities within the book. It is highly recommended that the supplies be purchased in advance to ensure that all materials are available for use.

We also recommend you read the "Before the Lesson," "During the Lesson," and "After the Lesson" instructions prior to completing the lesson with students. Written based on the experiences of other teachers, these tips will help you avoid pitfalls, maximize student learning and ensure that lessons run smoothly.

What if I Need Help Implementing These Lessons?

Our lessons are designed for teachers to easily implement them in their classroom. However, in the event that a teacher needs assistance, Stylish Schooling provides free support to teachers via email, phone, or our live support. Teachers are encouraged to visit our discussion boards online as well. Please visit our website, *www.stylishschooling.com*, for more information.

References

Arieli B. 2007. The integration of creative drama into science teaching. Unpublished PhD Thesis. Kansas State University.

Barthes, R. 1978. *Image-music-text*. S. Heath (Trans.) New York: Hill and Wang.

Bruner, J. 2004. Life as narrative. *Social Research* 71(3): 691–710.

Bruner, J. 1990. Culture and human development: A new look. *Human Development* 33(6): 344–355.

Craciun, D. 2010. Role-playing as a creative method in science education. *Journal of Science and Arts*. 1(12): 175–182.

Gee, J. P. 1991. What is literacy? In *Rewriting literacy: Culture and the discourse of the other*, ed. C. Mitchell and K. Weiler, 3–12. New York: Bergin and Garvey.

Gilbert, J., R. Hopkins, and G. Cooper. 2005. Faction or fiction: Using narrative pedagogy in school science education. Paper presented at the Redesigning Pedagogy: Research, Policy, and Practice conference, Singapore.

Graesser, A. C., M. Singer, and T. Trabasso. 1994. Constructing inferences during narrative text comprehension. *Psychological Review* 101: 371–395.

Howard, T. C. 2003. Culturally relevant pedagogy: Ingredients for critical teacher reflection." *Theory Into Practice* 42(3): 195–202.

Kirkland, K. 2003. Steppin' up and representin'. In *Becoming multicultural educators: Personal journey toward professional agency*, ed. G. Gay, 117–142. San

Francisco: Jossey-Bass.

Kreps-Frisch, J., and G. Saunders. 2008. Using stories in an introductory college biology course. *Journal of Biological Education* 42(4): 164–169.

Ladson-Billings, G., and W. Tate. 1995. Toward a critical race theory of education. *Teacher College Record* 97(1): 47–68.

McSharry, G., and S. Jones. 2000. Role-play in science teaching and learning. *School Science Review* 82 (298): 73–82.

Milne, C. 2011. Beyond argument in science: Science education as connected and separate knowing. In *Second International Handbook of Science Education*, ed. B. J. Fraser, K. McRobbie, and J. Campbell, 951–967. New York: Springer.

Varelas, M., C. C. Pappas, J. M. Kane, A. Arsenault, J. Hankes, and B. M. Cowan. 2008. Urban primary-grade children think and talk science: Curricular and instructional practices that nurture participation and argumentation. *Science Education* 92: 65–95.

Chapter 2
Aligning to Standards

Over the last few decades, standards have been proposed, modified, and implemented as a means for providing high expectations, common ground, and accountability for students and educators across districts and states. According to Learning First Alliance (2012), "when states and school districts set clear standards defining what students should know and be able to do, it focuses educational systems on priorities and actions to improve […] Teachers can understand what they need to teach and need to know" (under "The Promise of Standards-Based Reform"). This chapter will explicitly state how our book, *Diagnosis for Classroom Success: Making Anatomy and Physiology Come Alive*, aligns with both the Next Generation Science Standards (NGSS) and the Common Core State Standards (CCSS) so that science educators can easily express to administrators, parents, students, and fellow colleagues, how their work is meeting national guidelines.

The Next Generation Science Standards (NGSS)
Throughout the 2011–2012 school year, a team of 41 teachers from 26 states formed a committee with the goal of writing clear, concise, and practical national science standards for others in the science teaching profession to use in their classrooms. Although these standards are currently being drafted as this book is penned, the committee has already established three core elements:

- Dimension I: Practices
- Dimension II: Crosscutting Concepts
- Dimension III: Disciplinary Core Ideas

Below is not only a brief description of each element but specific examples as to how our book aligns with each one.

Chapter 2
Aligning to Standards

Dimension I: Practices

According to The National Academies Press's *A Framework for K–12 Science Education: Practices, Crosscutting Concepts, and Core Ideas* (2012), *practices* refers to the behaviors that scientists engage in as they investigate the natural world. This dimension encompasses eight key ideas:

Eight Key Ideas	NGSS Description	Our Book
Asking questions and defining problems	"Science begins with a question about a phenomenon, such as 'Why is the sky blue?' or 'What causes cancer?' and seeks to develop theories that can provide explanatory answers to such questions. A basic practice of the scientists formulating empirically answerable questions about phenomena, establishing what is already known, and determining what questions have yet to be satisfactorily answered" (p. 50).	Students will address the following four questions throughout this book and also develop their own questions as they go. • What is wrong with our four patients? • Why (and when) do these symptoms occur? • Who is most likely affected by these health conditions? • How can we prevent these conditions from happening?
Developing and using models	"Science involves the construction and use of a wide variety of models and simulations to help develop explanations about natural phenomena. Models make it possible to go beyond observables and imagine a world not yet seen. Models enable predictions of the form 'if . . . then . . . therefore' to be made in order to test hypothetical explanations" (p. 50).	The labs depicted in this book are designed to simulate the medical profession. The medical records provided will allow students to develop their initial hypothesis regarding the condition of each patient.
Planning and carrying out investigations	"Scientific investigation may be conducted in the field or the laboratory. A major practice of scientists is planning and carrying out a systematic investigation, which requires the identification of what is to be recorded and, if applicable, what are to be treated as the dependent and independent variables (control of variables). Observations and data collected from such work are used to test existing theories and explanations or to revise and develop new ones" (p. 50).	Students will complete a multitude of labs. Data collected from each patient will either support or negate their original hypotheses. • Lab 1: Urinalysis • Lab 2: Digestive By-Products and BMI Analysis • Lab 3: Blood Smears • Lab 4: HIV Test • Lab 5: Lung Capacity • Lab 6: Hormone Test • Lab 7: Performing Surgery

Chapter 2
Aligning to Standards

Eight Key Ideas	NGSS Description	Our Book
Analyzing and interpreting data	"Scientific investigations produce data that must be analyzed in order to derive meaning. Because data usually do not speak for themselves, scientists use a range of tools—including tabulation, graphical interpretation, visualization, and statistical analysis—to identify the significant features and patterns in the data. Sources of error are identified and the degree of certainty calculated. Modern technology makes the collection of large data sets much easier, thus providing many secondary sources for analysis" (p. 51).	During the "medical school" portion of this book, students will use an array of secondary sources to collect information on the four health conditions. For every lab, students will place data into a data table. During Labs 5 and 6, students will analyze a graph of lung capacity and a graph of glucose tolerance test results. As labs are completed, students will continuously refer to their "medical school" research to help make an appropriate diagnosis.
Using mathematics and computational thinking	"In science, mathematics and computation are fundamental tools for representing physical variables and their relationships. They are used for a range of tasks, such as constructing simulations, statistically analyzing data, and recognizing, expressing, and applying quantitative relationships. Mathematical and computational approaches enable predictions of the behavior of physical systems, along with the testing of such predictions. Moreover, statistical techniques are invaluable for assessing the significance of patterns or correlations" (p. 51).	Students will perform a set of algebraic equations during Lab 6 when comparing their patients' estimated lung capacity with that of their expected lung capacity.
Constructing explanations	"The goal of science is the construction of theories that can provide explanatory accounts of features of the world. A theory becomes accepted when it has been shown to be superior to other explanations in the breadth of phenomena it accounts for and in its explanatory coherence and parsimony. Scientific explanations are explicit applications of theory to a specific situation of phenomenon, perhaps with the intermediary of theory-based model for the system under study. The goal for students is to construct logically coherent explanations of phenomena that incorporate their current understanding of science, or a model that represents it, and are consistent with the available evidence" (p. 52).	Upon completing the exercises in this book, students should be able to explain *how* one's behaviors and hereditary makeup may result in one of the four health conditions, in addition to explaining *why* the various symptoms come about as a result of each health condition. At the end of this book, students will share, in writing, an explanation of various science phenomena.

Chapter 2
Aligning to Standards

Eight Key Ideas	NGSS Description	Our Book
Engaging in argument from evidence	"In science, reasoning and argument are essential for identifying the strengths and weaknesses of a line of reasoning and for finding the best explanation for a natural phenomenon. Scientists must defend their explanations, formulate evidence based on a solid foundation of data, examine their own understanding in light of the evidence and comments offered by others, and collaborate with peers in searching for the best explanation for the phenomenon being investigated" (p.52).	Throughout this book, students will collaborate with their peers on a quest to determine what is wrong with their patients. After analyzing all the data, students are expected to make a diagnosis and defend it based on evidence.
Obtaining, evaluating, and communicating information	Science cannot advance if scientists are unable to communicate their findings clearly and persuasively or to learn about the findings of others. A major practice of science is thus the communication of ideas and the results of inquiry—orally, in writing, with the use of tables, diagrams, graphs, and equations, and by engaging in extended discussions with scientific peers. Science requires the ability to derive meaning from scientific texts (such as papers, the internet, symposia, and lectures), to evaluate the scientific validity of the information thus acquired, and to integrate that information" (p. 53).	At the end of this book, students will share their findings with others in an oral presentation format. It is suggested that teachers provide an authentic audience for their students by inviting doctors, medical school students, or other science teachers to participate in the discussion. The presentation format requires that students include data tables, graphs, and equations to support their work.

Dimension II: Crosscutting Concepts

Crosscutting Concepts are core ideas that connect the different domains of science (Physical Science, Life Science, Earth and Space Science, and Engineering/Technology/Application of Science) together. Again, The National Academies Press's *A Framework for K–12 Science Education: Practices, Crosscutting Concepts, and Core Ideas* (2012) have identified seven key ideas:

Chapter 2
Aligning to Standards

Seven Key Ideas	NGSS Description	Our Book
Patterns, similarity, and diversity	"Observed patterns of forms and events guide organization and classification, and they prompt questions about relationships and the factors that influence them" (p. 84).	When doctors diagnose their patients, they look for patterns in their patients' medical records (i.e., family history, previous health conditions, and so on), current symptoms, and lab results. The same will be expected of students using this book.
Cause and effect: Mechanism and explanation	"Events have causes, sometimes simple, sometimes multifaceted. A major activity of science is investigating and explaining causal relationships and the mechanisms by which they are mediated. Such mechanisms can then be tested across given contexts and used to predict and explain events in new contexts" (p. 84).	By analyzing the four patients' medical records for various risk factors (i.e., the cause), and observing each patients' symptoms and results from the first six laboratories, students will determine which condition (i.e., the effect) is present in the patients. Additionally, students will need to determine why the symptoms occur. Finally, the analysis questions, provided at the end of each lab, will require students to predict or explain events in a new context.
Scale, proportion, and quantity	"In considering phenomena, it is critical to recognize what is relevant at different measures of size, time and energy and to recognize how changes in scale, proportion or quantity affect a system's structure or performance" (p. 84).	Lab 7, the rat dissection, addresses the notion of scale and proportion best. Rats, like humans, are mammals; therefore, their anatomy and physiology parallels our own. In this lab, students will observe the vast size of the liver (taking up a large proportion of the rat's abdomen cavity), the length of the small intestine surpassing the length of the rat's body, and make several other observations regarding the structure and function of organs.
Systems and system models	"Defining the system under study—specifying its boundaries and making explicit a model of that system—provides tools for understanding and testing ideas that are applicable throughout science and engineering" (p. 84).	Throughout this book there are seven body systems studied: the excretory, digestive, circulatory, immune, respiratory, reproductive, and endocrine. Students will first study these systems separately, but as they make their way through the various labs, it will become evident that these systems interact and depend on one another.
Energy and matter	"Tracking fluxes of energy and matter into, out of, and within systems helps one understand the systems' possibilities and limitations" (p. 84).	Within this book, the opportunity to discuss energy occurs during Lab 2 and Lab 6, when discussing nutrition and glucose levels in the blood.
Structure and function	"The way in which an object or living thing is shaped and its substructure determine many of its properties and functions" (p. 84).	As students progress through this book, they are consistently learning about how the organs of the body appear and function. When students perform Lab 7, they will be required to identify major organs of the body.

DIAGNOSIS FOR CLASSROOM SUCCESS: Making Anatomy & Physiology Come Alive

Chapter 2
Aligning to Standards

Seven Key Ideas	NGSS Description	Our Book
Stability and change	"For natural and build systems alike, conditions of stability and determinants of rates of change or evolution of a system are critical elements of study" (p. 84).	Although each organ of the body serves a unique purpose in keeping us alive and well, there are various reasons homeostasis can be disrupted. Students will learn about genetics and behavior (i.e., engaging in risky sexual behavior, poor diet, and lack of exercise).

Having experienced all seven themes within the context of our book, students should, expectantly, be able to apply these skills in new contexts, be it in other units, science disciplines or that of other content areas.

Dimension III: Disciplinary Core Ideas

Disciplinary Core Ideas allow teachers to focus their curricula, pedagogy, and assessments on the four most important elements of science, as identified by the NGSS. The NGSS recommend that a science unit meet two of the following four criteria:

Four Core Ideas	Our Book
Broad importance across all sciences or key organizational concepts within a single discipline	Teachers will find themselves referring back to the content within this book when teaching other units within their discipline (i.e., cells, genetics, evolution).
Key tool for understanding or illustrating more complex ideas and solving problems	Students will be following a realistic, collaborative approach to solving problems.
Relating to the interests and life experiences of students, societal and personal concerns	The four health conditions (sickle cell anemia, HIV, pregnancy, and diabetes) are extremely relevant for students, and equally concerning in today's society. By tapping into cultural relevancy, students will be enthusiastic about learning the content as well.
Be teachable and learnable over multiple grades at increasing levels of depth and sophistication	This book is designed for grades 7–12, to supplement the current teaching pedagogy used by teachers. Teachers can differentiate the book according to the text provided during the "medical school" research component, and for more advanced students, teachers can incorporate two additional diseases into the mix by downloading our supplemental worksheet on our website.

Chapter 2
Aligning to Standards

The Common Core State Standards

In addition to the Next Generation Science Standards, this book aligns with the Common Core English Language Arts standards for Science and Technical Subjects and for Writing. The Common Core State Standards was recently introduced, across the country, as a means for preparing students for college and beyond. According to The Common Core State Standards (2012) website, the standards were coordinated by the National Governors Association Center for Best Practices (NGA Center) and the Council of Chief State School Officers (CCSSO) and derived by teachers, school administrators and experts.

The Common Core Literacy Standards for Science and Technical Subjects is divided into four categories: Key Ideas and Details, Craft and Structure, Integration of Knowledge and Ideas, and Range of Reading and Level of Text Complexity. In total, there are 10 standards. Our book taps into seven of them.

The Common Core	Our Book
CCSS.ELA–Literacy.RST.9-10.1 Cite specific textual evidence to support analysis of science and technical texts, attending to the precise details of explanations or descriptions. **CCSS.ELA–Literacy.RST.9-10.2** Determine the central ideas or conclusions of a text; trace the text's explanation or depiction of a complex process, phenomenon, or concept; provide an accurate summary of the text. **CCSS.ELA–Literacy.RST.9-10.10** By the end of grade 10, read and comprehend science/technical texts in the grades 9–10 text complexity band independently and proficiently.	During the "medical school" research chapter of our book, students are required to answer questions related to four health conditions (diabetes, HIV, pregnancy, and sickle cell anemia). During this task, students should be able to demonstrate their ability to determine central ideas of complex, yet age appropriate, texts. Likewise, they should be able to paraphrase or cite the text properly. Additionally, teachers should use their judgment when deciding on whether or not the research task should be completed independently or in groups. We suggest "jigsawing" students. Jigsawing requires teachers to separate students into "expert" groups. Each group, however, is responsible for a different topic (i.e., one group would be the expert on sickle cell anemia, while another group would be an expert on HIV). After the allotted time, one expert from each group combines to form a new group of mixed expertise. Each group member is expected to share their wealth of knowledge with his or her new group members. In our book, we have modified this activity so that each expert must learn his or her assigned topic independently and then come together with the group for collaboration. This modification appears to hold students accountable for their learning and puts pressure on them to come to the group with accurate and complete work.

Chapter 2
Aligning to Standards

CCSS.ELA–Litearcy.RST.9-10.3 Follow precisely a complex multistep procedure when carrying out experiments, taking measurements, or performing technical tasks, attending to special cases or exceptions defined in the text.	The seven labs portrayed throughout this book provide a multitude of opportunities for students to demonstrate their ability to follow multistep procedures. Lab 6, in particular, entails quite a bit of measuring and calculating on the students' part.
CCSS.ELA–Litearcy.RST.9-10.4 Determine the meaning of symbols, key terms, and other domain-specific words and phrases as they are used in a specific scientific or technical context relevant to grades 9–10 texts and topics. **CCSS.ELA–Litearcy.RST.9-10.5** Analyze the structure of the relationships among concepts in a text, including relationships among key terms (e.g., *force, friction, reaction force, energy*).	This book offers several "Before the Lesson" key terms and domain-specific words and phrases for teachers to introduce, and use, within their classroom. Since the body systems, their diseases, and their corresponding labs are consistently interconnected and explicitly discussed throughout this book, students will inevitably make connections between concepts and key terms as they move through it.
CCSS.ELA–Litearcy.RST.9-10.7 Translate quantitative or technical information expressed in words in a text into visual form (e.g., a table or chart) and translate information expressed visually or mathematically (e.g., in an equation) into words.	As students complete each lab, they are expected to record and input their data into tables. Likewise, within this book there are several charts, graphs, and equations, for students to interpret. Some include: Lab 2's BMI chart, Lab 5's lung capacity graph and equations, and Lab 6's glucose tolerance test results.

Similar to the Science and Technical Subject standards, there are also several Common Core English Language Arts standards for Writing that correspond to this book. Within the Writing section, there are four categories as well: Text Types and Purposes, Production and Distribution of Writing, Research to Build and Present, and Range of Writing, totaling 10 standards. This book taps into 5 of those 10.

The Common Core	This Book
CCSS.ELA–Literacy.WHST.9-10.1 Write arguments focused on discipline-specific content. **CCSS.ELA–Literacy.WHST.9-10.2** Write informative/explanatory texts, including the narration of historical events, scientific procedures/experiments, or technical processes. **CCSS.ELA–Literacy.WHST.9-10.6** Use technology, including the Internet, to produce, publish, and update individual or shared writing products, taking advantage of technology's capacity to link to other information and to display information flexibly and dynamically. **CCSS.ELA–Literacy.WHST.9-10.5** Develop and strengthen writing as needed by planning, revising, editing, rewriting, or trying a new approach, focusing on addressing what is most significant for a specific purpose and audience. **CCSS.ELA–Literacy.WHST.9-10.4** Produce clear and coherent writing in which the development, organization, and style are appropriate to task, purpose, and audience.	When evaluating your students' knowledge, there are multiple options provided in this book. One of those options entails a written component, in which students are expected to produce clear and coherent writing. Similarly, the writing should contain arguments that support claims and provide valid reasoning, as well as relevant and sufficient evidence from experiments performed. Since students will typically be working in groups of three or four, it is highly advised that students use technology (such as Google Drive or Dropbox) to share and collaborate their documents. For the presentation component of this project, students should also use technology that facilitates collaboration among group members and also enables the teacher to provide feedback that can allow for multiple revisions. Lastly, since it is suggested that students present their work to medical professionals, science educators, and/or other students (i.e., via a roundtable discussion), students should be expected to maintain a sense of professionalism intended for their specific audience.

References

Learning First Alliance (n.d.). *Standards and accountability: A call by the Learning First Alliance for mid-course corrections.* www.learningfirst.org/news/standards/ full.html

National Governors Association Center for Best Practices, Council of Chief State School Officers. 2010. *Common core state standards.* Washington, DC: National Governors Association Center for Best Practices, Council of Chief State School Officers. *www.corestandards.org/about-the-standards*

National Governors Association Center for Best Practices, Council of Chief State School Officers. (2010). *Common Core State Standards (ELA–Literacy).* Washington, DC: National Governors Association Center for Best Practices, Council of Chief State School Officers. *www.corestandards.org/ELA-Literacy/RST/9-10*

National Research Council (NRC). forthcoming. *Next Generation Science Standards.* www.nextgenscience.org/three-dimensions

National Research Council. 2012. *A framework for K–12 science education: Practices, crosscutting concepts, and core ideas.* Washington, DC: National Academies Press.

Chapter 3
Teacher Prep Made Easy

Materials

For your convenience, we have included the following charts to simplify prep time.

Shaded regions = One-time purchase only! Reusable each year.

Chapter 3
Teacher Prep Made Easy

Lab 1: Urinalysis						
Number needed per group	Item Name	Vendor	Catalog #	Cost	Number I Need to Order	Total Cost
8	Test tubes					
1	Test tube rack					
1	Test tube clamp					
1	10 ml graduated cylinder					
20 ml	500 ml Benedict's solution					
20 ml	500 ml Biuret reagent					
1	250 ml beaker					
2	Droppers					
1 per person	Goggles					
1	Hot plate					
2	Paper cups					
N/A	Bottle of apple juice					
N/A	1 container of egg whites					
N/A	Yellow food coloring					
N/A	Supply of water					

Total Cost for This Lab: _____

Lab 2: Digestive By-Products and BMI Analysis						
Number Needed per group	Item Name	Vendor	Catalog #	Cost	Number I Need to Order	Total Cost
6	Test tubes					
1	Test tube rack					
1	Test tube clamp					
1	10 ml graduated cylinder					
20 ml	500 ml Benedict's solution					
20 ml	500 ml Biuret reagent					
20 ml	Lugol solution (Iodine)					
1	Hot plate, economy, single burner					
1	250 ml beaker					
2	Droppers					
1 per person	Goggles					
3	Paper cups					
N/A	Grapes					
N/A	1 banana					
N/A	Bottle of apple juice					
N/A	Can of chickpeas					
N/A	Carton of cottage cheese					

Total Cost for This Lab: _____

Chapter 3
Teacher Prep Made Easy

Lab 3: Blood Smears						
Number Needed per Group	Item Name	Vendor	Catalog #	Cost	Number I Need to Order	Total Cost
3	Human blood smear (Wright's stain)					
1	Human sickle cell anemia smear					
1	Compound microscope					

Total Cost for This Lab: _____

Lab 4: HIV Test						
Number Needed per Group	Item Name	Vendor	Catalog #	Cost	Number I Need to Order	Total Cost
4 per person	Paper cups					
4 per group	Test tubes					
1	Test tube rack					
1	Dropper					
1 per person	Goggles					
N/A	500 ml sodium hydroxide, 50% solution in water					
N/A	100 ml of 1% phenol-phthalein					
N/A	Distilled water					

Total Cost for This Lab: _____

Chapter 3
Teacher Prep Made Easy

Lab 5: Lung Capacity						
Number Needed per Group	Item Name	Vendor	Catalog #	Cost	Number I Need to Order	Total Cost
1 per person	Scientific calculators					
1	Wind-up measuring tape					
4	Balloons					

Total Cost for This Lab: _____

Lab 6: Hormone						
Number needed per group	Item Name	Vendor	Catalog #	Cost	Number I Need to Order	Total Cost
2 strips	Blue litmus paper					
2	Paper cups					
N/A	Bottled water					
N/A	Yellow food coloring					
N/A	Bottle of vinegar					

Total Cost for This Lab: _____

Lab 7: Performing Surgery						
Number needed per group	Item Name	Vendor	Catalog #	Cost	Number I Need to Order	Total Cost
1	Standard dissecting tray					
1	Student dissecting set					
1	Perfect solution white rat, plain					

Total Cost for This Lab: _____

Chapter 3
Teacher Prep Made Easy

Pacing Chart

Whether your district schedules 45-minute periods or 90-minute blocks, our curricula can be applied to fit your scheduling needs. Below is our recommended time frame for the activities within the student edition (Note: Student note-taking and lectures not included).

Recommended Time[1]	Student Edition
180 minutes	Chapter 1—Medical School Research
45 minutes	Chapter 2—Medical Records
45 minutes	Chapter 3—Urinalysis
60 minutes	Chapter 4—Digestive By-Products and BMI Analysis
45 minutes	Chapter 5—Blood Smears
30 minutes	Chapter 6—HIV Test
60 minutes	Chapter 7—Lung Capacity
30 minutes	Chapter 8—Hormone Test
45-60 minutes	Chapter 9—Performing Surgery
TBD by teacher	Chapters 10 and 11—Prognosis and Doctor (Student) Evaluation 360 minutes to develop[2] and revise PowerPoint Roundtable presentations take approximately 45–60 minutes to complete per group

1. Please note that students of varying cognitive levels may need different time frames. We suggest adjusting accordingly.
2. If class time is limited, we recommend assigning the PowerPoint as homework.

Role-Play Comparison Chart

As discussed in Chapter 1, role-play makes up a large percentage of our curricula. Before beginning any of our workbooks, it is important to clarify to your students that these laboratories are merely simulations. We suggest that science educators use the chart below to help students differentiate between our book's procedures and that of the medical profession.

	Medical Procedure	**Our Book**
Lab 1: Urinalysis	When patients provide a urine sample, it is sent to a laboratory and three tests are performed: visual, chemical, and microscopic. • Visually, doctors will look for features similar to that of your students: color and transparency. • Some chemical tests performed by doctors, in addition to glucose and protein, are solute concentrations, pH levels, ketones, blood, nitrites, and so forth. To test for the presence of sugar, however, a glucose strip is dipped into the urine sample, indicating the level of sugar present. Protein is tested in a similar fashion. • Microscopic examinations are typically not done unless there is cause for concern (i.e., blood in the urine).	In our book, students observe three physical features of the simulated urine samples: color, odor and transparency, and two chemical features (the presence of sugar and protein). The chemical tests performed in this book (Benedict and Biuret), although completely accurate, are not the tests doctors would use in their office.
Lab 2: Digestive By-Products and BMI Analysis	In contrast to our book, doctors do not perform nutrient tests on the digestive by-products of patients suffering from excessive vomiting or diarrhea. They do, however, try to avoid dehydration by replacing lost nutrients with intravenous fluids and must be conscious of the glycemic index (GI) when arranging meal plans for diabetic patients. As for utilizing BMI charts, doctors understand the limitations (i.e., the weight of muscle mass) that these charts have in determining whether or not a person is healthy internally. For instance, men and women are not always assessed using the same BMI chart. Nor should children be assessed using the same BMI chart as adults. BMI charts can, however, be used in combination with other diagnostic tools (i.e., urinalysis, blood smears, glucose tolerance tests) for determining a range of health conditions.	In our book, the chemical tests performed on the digestive by-products of patients #2 and #3 (Benedict, Biuret, and Lugol) are all legitimate tests for detecting glucose, protein, and starch, respectively. However, as indicated, doctors have no need for using these tests in their workplace. The BMI chart in this book is specifically for adults considering the ages of the four patients. However, BMI charts have been modified for children and different genders.

Chapter 3
Teacher Prep Made Easy

	Medical Procedure	**Our Book**
Lab 3: Blood Smears	Sickle cell anemia is screened when babies are born. Because sickle cell anemia predominantly affects African Americans, they are the population most commonly tested for it, specifically when both parents carry the gene. To test for sickle cell anemia, blood samples are drawn and checked for sickled hemoglobin. If a baby tests positive for the sickled hemoglobin, a second blood test is processed to confirm the diagnosis. Another option is to perform a DNA test by checking the amniotic fluid of the growing baby for copies of the sickle cell gene. Lastly, as in the case of our book, patients may also consider a blood smear and a complete blood count to look for abnormalities in a red blood cell's shape or levels.	In our story, Patient #4 is an African American adult, and is therefore screened for sickle cell at a later age than would be typical.
Lab 4: HIV Test	When an individual wants to determine if he or she is HIV positive, the person can submit one of two body fluids: blood or saliva. There are several techniques used to determine if a person is infected with HIV. Two of these techniques look to see if a person is producing antibodies against the virus. • The Enzyme-linked Immunosorbent Assay (ELISA) test is the first one used to detect HIV infections. If the ELISA results are positive, a second one is used to confirm the results. • The Western blot test is considered more difficult to perform and is typically used after ELISA tests produces two positive results. Contrastingly, polymerase chain reactions (PCR), also known as the viral load test, do not look for antibodies against HIV. PCR tests seek out the genetic material of HIV or the white blood cells infected by HIV. Since PCR tests require intense training and expensive equipment, it is less commonly used. However, it is far more sensitive to detecting the virus.	When students perform the HIV test in this book, they are utilizing body fluids from their patients and testing for antibodies. Though, where these body fluids came from (blood or saliva) is not specified. The HIV test in this book is simply a chemical change that occurs when phenolphthalein (and indicator) is in the presence of a base (NaOH) turning the solution from clear to pink. This chemical process takes less than one second to occur, which is not typical of an HIV test.

Chapter 3
Teacher Prep Made Easy

	Medical Procedure	**Our Book**
Lab 5: Lung Capacity	In order to determine if a patient's lungs are being compromised, a doctor will use a tool called a spirometer. A spirometer contains a tube for patients to blow into. The patient may be asked to alter his or her breathing (i.e., fast, slow, or deep). A brief description of a spirometer is mentioned in Chapter 7 of the student edition. When checking for lung volume, doctors will also use a machine called a plethysmograph, This procedure entails patients sitting in a booth, breathing in a tube, and doctor's measuring a change in pressure within the booth, to determine lung volume. Finally, if a doctor wants to determine how much dissolved oxygen is in his/her patient's bloodstream, he or she may request a pulse oximetry test, which requires the patient to attach a sensor to his or her ear or finger.	Considering the difficulty in, and expense of, purchasing classroom spirometers, this book provides an alternative means for estimating one's actual and acceptable lung capacities. Both are perfectly good tools for determining our patients' lung capacities since they factor in height, weight, and gender. However, spirometers provide less room for error since they do not require the doctor to perform measurements and calculations.
Lab 6: Hormone Test	When a woman misses her first menstrual cycle, it is usually a sign that she is pregnant. This most often results in her purchasing an at-home pregnancy test. If the test reads positive, she is producing the hormone hCG. Most women, however, will also schedule an appointment with their gynecologist for additional blood work to verify the presence of this hormone. It should be noted, however, that hCG is typically tested for in a regular urinalysis. This lab also tests for insulin. A description of a glucose tolerance test is found in Chapter 8 of the student edition.	As mentioned already, all bodily fluids are simulated in this book. Since a placenta produces the hormone, hCG, it is nearly impossible to replicate it in the classroom. As a result, patients #1 and #2's urine samples have been substituted with water and yellow food coloring, hCG with vinegar, and a pregnancy test with litmus paper. The glucose tolerance test does not provide simulated hands-on experience for students. However, the graph produced is a true depiction of what levels would look like for diabetics and non-diabetics at various time intervals.

References

American Association for Clinical Chemistry. Lab tests online—Sickle cell tests. *http://labtestsonline.org/understanding/analytes/sickle/tab/test*

American Association for Clinical Chemistry. Lab tests online—Urinalysis. *http://labtestsonline.org/understanding/analytes/urinalysis/tab/test*

Avert. HIV testing. *www.avert.org/testing.htm*

National Heart Lung and Blood Institute. How is sickle cell anemia diagnosed? *www.nhlbi.nih.gov/health/health-topics/topics/sca/diagnosis.html*

National Heart, Lung, and Blood Institute. What to expect during lung function tests. *www.nhlbi.nih.gov/health/health-topics/topics/lft/during.html*

WebMD. HIV and AIDS health center: Human immunodeficiency virus. *www.webmd.com/hiv-aids/human-immunodeficiency-virus-hiv-test*

2 DURING

1
2
3

STUDENT EDITION

Chapter 4
Teacher Guide for Earning Your White Coats
Medical School Research

Before the Lesson

- Be energetic! Teaching, to some extent, is an acting job. The more enthusiastic you are about teaching this curriculum, the more enthusiastic your students will be about learning it.

- Organize your students into groups of four. It is highly recommended that groups are arranged heterogeneously by skill level.

- Make sure you have access to computers with Internet. If not, print out age-appropriate readings about sickle cell anemia, HIV, diabetes and pregnancy. Highly recommended websites include:

    - Pub Med Health (*www.ncbi.nlm.nih.gov/pubmedhealth*)
    - Mayo Clinic (*www.mayoclinic.com*)
    - Kids Health (*http://kidshealth.org*)

During the Lesson

- Read the overview of tasks to familiarize students with the overarching goal of the project.

- We recommend you "jigsaw" your students for the research component. Assign one student in each group to a different health condition or have them select which one they would like to do on their own. Students should complete their assigned table: Table 1.1, 1.2, 1.3, or 1.4 (pp. 35–38, teacher edition). Once completed, they should communicate their findings to their group members.

- Once students have completed the research portion, sign their worksheets to ensure all research components are completed thoroughly and accurately.

Chapter 4
Teacher Guide for Earning Your White Coats: Medical School Research

- Read the Hippocratic oath as an entire class. Ensure that students annotate as they read. Have a discussion about the Hippocratic oath. Once done, have them sign it. Aside from the discussion questions mentioned below, this is also a great time to discuss controversial topics such as euthanasia, abortion, celebrity deaths, and access to prescription drugs for recreational use. Discussion questions include:
    - What questions does this reading passage raise for you?
    - What are the advantages of having doctors follow the Hippocratic oath?
    - What are some disadvantages or concerns doctors or patients might have regarding the Hippocratic oath?
    - What suggestions do you have for revising the Hippocratic oath?
    - What is your stance on it?
- Students should record their responses to the discussion questions in Table 1.5 (p. 41).

After the Lesson

- Discuss what would happen after a student graduates from medical school. Topics to consider:
    - Taking the Boards
    - Applying for residency programs

Earning Your White Coats: Medical School Research

Task Overview

To successfully complete the Anatomy and Physiology unit, you and your classmates will be required to:

1. *Attend and graduate medical school:* In order to graduate medical school, all students must complete the research portion of this project. All medical school students will be required to investigate, as thoroughly as possible, the causes of, symptoms of, and potential treatments for four health conditions. Since there will be limited time to complete this task, working efficiently as a group will be critical. Once the research portion is approved, permission to graduate will be granted by the medical school president (your teacher).

2. *Sign the Hippocratic oath:* Graduating medical students will be required to read and sign the Hippocratic oath before accepting and treating patients, ensuring all soon-to-be doctors understand the role of ethics in medicine.

3. *Meet your patients:* Based on both the knowledge obtained from medical school and the medical records provided by the four patients, doctors (you the students) will develop an initial hypothesis.

4. *Run diagnostic tests on patients:* Doctors will conduct six labs to help diagnose the four patients. Lab 7 will not assist in the diagnosis of *your* four primary patients.

   - Lab 1: Urinalysis
   - Lab 2: Digestive By-Products and BMI Analysis
   - Lab 3: Blood Smears
   - Lab 4: HIV Test
   - Lab 5: Lung Capacity
   - Lab 6: Hormone Test
   - Lab 7: Performing Surgery

5. *Diagnose patients and develop a prognosis:* Once the group reaches a consensus regarding each patient's appropriate diagnosis, the medical chart must be completed so that (a) all patients can thoroughly understand their prognosis and (b) a prescribed treatment can be filled by a pharmacist.

6. *Develop a written, visual, and/or oral report:* All doctors will be evaluated on their ability to collect and analyze *evidence*, their ability to make *con-*

Chapter 4
Teacher Guide for Earning Your White Coats: Medical School Research

Answer Key

nections between the biology content and the various laboratories used to diagnose patients, and on their understanding of the topics discussed.

7. *Receive feedback from your evaluators.* Evaluators will determine whether or not a doctor may continue practicing medicine (and has therefore passed) or if a doctor is at risk of losing his or her license (and is therefore not familiar enough with the content).

Welcome to Medical School!

Greetings! Your professor has assigned you and a team of medical students to conduct research on the following four health conditions: sickle cell anemia, the human immunodeficiency virus (HIV), pregnancy, and diabetes. Within your team's research, be sure to include the causes of, symptoms of, and treatments of (if any exist) the aforementioned conditions. Your professor (classroom teacher) will determine the time allotted to complete this task. Remember, medical school requires dedication, hard work, and great attention to detail. Stay focused and good luck!

Study Group at the Library

You and your team have headed straight to the medical school library. As a group, you decide it is best to split up research tasks and share your findings afterward. Before starting, each group member selects one of the four health conditions to study. Everyone in the group promises to complete the table (Tables 1.1, 1.2, 1.3, and 1.4, pp. 4–7, student edition) that corresponds with the assigned condition. When everyone has finished, be sure to communicate your findings to others so that they, too, learn about the condition.

Study Group Assignments (Student Name)

1. _____ will study sickle cell anemia and complete Table 1.1.

2. _____ will study human immunodeficiency virus and complete Table 1.2.

3. _____ will study pregnancy and complete Table 1.3.

4. _____ will study diabetes and complete Table 1.4.

Answer Key

Chapter 4
Teacher Guide for Earning Your White Coats: Medical School Research

TABLE 1.1. INFORMATION COLLECTED ON SICKLE CELL ANEMIA

Sickle Cell Anemia	
Causes 1. How does one get sickle cell anemia?	Sickle cell anemia occurs in individuals who inherit two copies of the sickle cell gene (HS) from parents.
2. How does one get the sickle cell trait?	Sickle cell trait occurs in individuals who inherit one copy of the sickle cell genes from a parent.
Symptoms 1. What are the symptoms of sickle cell anemia?	Symptoms of sickle cell anemia include: jaundice (yellowing of skin and eyes), blood clots and strokes, fatigue, and joint pain.
2. What cells are affected by sickle cell?	Sickle cell anemia affects the red blood cells.
3. What shape do these cells turn into?	Red blood cells (normally a circular donut shape) turn into a sickle (crescent-moon) shape.
4. What protein is mutated on these cells? Explain how this is related to symptoms of sickle cell anemia.	Hemoglobin is the protein that is mutated on a red blood cell. Hemoglobin is responsible for carrying four oxygen molecules. When the hemoglobin protein is mutated, the red blood cell becomes deficient in oxygen, carrying only two molecules. This can cause shortness of breath. Additionally, the mutated hemoglobin is what causes the shape of a red blood cell to change. As a result, red blood cells cannot move through blood vessels as easily, causing blood clots and pain.
Treatment/cures 1. What cures exist?	Sickle cell is not curable and can potentially be fatal due to infections or organ failure.
2. What treatments exist?	Several treatments exist: folic acid supplements, penicillin and other antibiotics (to deal with infections), blood transfusions, hydroxyurea (medicine), kidney dialysis (due to complications of sickle cell), and so on.

Chapter 4
Teacher Guide for Earning Your White Coats: Medical School Research

Answer Key

TABLE 1.2. INFORMATION COLLECTED ON HUMAN IMMUNODEFICIENCY VIRUS (HIV)

Human Immunodeficiency Virus (HIV)	
Causes 1. How is HIV contracted?	HIV is contracted through bodily fluids: semen/vaginal fluid (unprotected sex), blood to blood contact (needles), mother to child (breastfeeding, birth).
2. Is HIV caused by a virus or bacteria?	HIV is a virus.
3. What type of cell does HIV attack in the immune system?	HIV attacks white blood cells in the immune system and specifically, the T-cells.
Symptoms 1. List the symptoms at different stages of an HIV infection.	(a) Directly after infection: HIV is undetectable. (b) 3-6 months after infection: Antibodies form in the bloodstream (detectable by an HIV test), fever, sore throats, aching muscles and joints, and swollen glands. (c) Years after the infection: White blood cells continue to drop (below 200 T-cells = full blown Acquired Immunodeficiency Syndrome), rashes, diarrhea, fatigue, weight loss, more susceptible to opportunistic infections such as: pneumonia, tuberculosis (TB), candidiasis thrush (fungal infection), and so on.
Treatment/cures 1. What cures exist? 2. What treatments exist?	Like all viral infections, HIV cannot be cured. HIV treatments do exist: Highly Active Anti-Retroviral Therapy (HAART) works by preventing the replication of HIV (i.e., blocking CD-4 receptors on T-cells).

Answer Key

Chapter 4
Teacher Guide for Earning Your White Coats: Medical School Research

TABLE 1.3. INFORMATION COLLECTED ON PREGNANCY

Pregnancy	
Causes	
1. What is fertilization?	Fertilization occurs when a sperm cell fuses with an egg cell.
2. How does fertilization take place?	In mammals (i.e., humans), birds, and reptiles, fertilization occurs internally. Sperm must be ejaculated into the female's vaginal canal, make way through the cervix and uterus, until it reaches the egg in the fallopian tube.
3. Where in the female does the egg get fertilized?	Fertilization occurs in the fallopian tube.
4. Where does the female egg travel upon fertilization?	Assuming everything goes according to plan, once the egg is fertilized, it travels to the uterus and will implant itself into the uterine lining.
Symptoms	
1. What are the symptoms of pregnancy?	(a) What hormone is released by the placenta and detected by a pregnancy test? Human Chorionic Gonadotropin (hCG) What are the symptoms of the (b) first trimester? Morning sickness, food cravings, heightened sense of smell, frequent urination, constipation, heartburn, breast tenderness (c) second trimester? Larger breasts (milk production), baby grows 3-4 pounds a month, skin changes, stretch marks, nasal and gum problems, leg cramps, shortness of breath, vaginal discharge, Braxton Hicks contractions, and so on (d) third trimester? Continued breast growth, more weight gain, shortness of breath, heartburn, swelling, frequent urination, various veins, backaches, Braxton Hicks contractions, and so on.
Treatment/cures	
1. What methods of birth control exist? Describe them.	There are several types of birth control methods: abstinence, implant, patch, pills, shots, sponges, rings, cervical caps, condoms, diaphragm, IUD, spermicides, vasectomy, sterilization in women.
2. What preventative health measures are recommended for an expecting mother?	When women are pregnant, they should avoid drugs and alcohol to prevent developmental problems of the baby. They should also exercise and eat right. Women are often recommended to take pre-natal vitamins (folic acid, calcium, iron).

Chapter 4
Teacher Guide for Earning Your White Coats: Medical School Research

Answer Key

TABLE 1.4. INFORMATION COLLECTED ON DIABETES

Diabetes	
Causes	
1. What is diabetes?	Diabetes is a disease in which a person does not produce, or properly use, insulin, resulting in high glucose levels in the blood.
2. What organ does not function properly in a diabetic?	The organ that does not function properly in a diabetic is the pancreas.
3. What is insulin?	Insulin is the hormone that is produced by the pancreas. It helps lower blood sugar levels by allowing the glucose to enter cells.
4. How does someone obtain type 1 diabetes?	Type 1 diabetes is typically called juvenile diabetes because it is genetic (most likely an autoimmune disorder of pancreatic cells called the Islets of Langerhans) and found in young children. The pancreas does not produce insulin.
5. How might someone develop type 2 diabetes?	Type 2 diabetes is typically found in young adults or the elderly. The insulin produced by the pancreas is no longer efficient at lowering blood sugar levels. Type 2 is often brought on by poor diet and lack of exercise and often occurs in overweight and obese individuals.
Symptoms	
1. What symptoms will a diabetic experience?	Blurry vision, dry skin, frequent thirst and urination, fatigue, hunger, weight loss (type 1), and overweight (type 2), and later on one might experience nerve damage and kidney damage.
Treatment/cures	
1. How does a person manage type 1 diabetes?	There is no cure for diabetes, but people can manage it through meal planning (diet), exercise, monitoring blood glucose levels, insulin injections, and so on.
2. How does a person manage type 2 diabetes?	(Same answer for #2)

> Answer Key

Chapter 4
Teacher Guide for Earning Your White Coats: Medical School Research

Medical School President's Signature of Approval:

_____

Graduation

Congratulations! You may *officially* graduate medical school! As a graduating medical school class, you must read and sign the modernized version of the Hippocratic oath. According to the Public Broadcasting Station's website, the Hippocratic oath originated in fifth century BC as a means of protecting patients with a code of ethics to be followed by health care professionals and physicians. Due to the outdated nature of the original Hippocratic oath, with its references to gods, goddesses, and slavery, it was revised in 1964.

As you read the modernized version of the Hippocratic oath aloud, annotate the text. Place a question mark next to sentences that puzzle you, a "W" next to sentences that worry or concern you, and an "E" next to sentences that excite or seem beneficial to you.

Chapter 4
Teacher Guide for Earning Your White Coats: Medical School Research

Answer Key

The Hippocratic Oath (The Modern Version)

I swear to fulfill, to the best of my ability and judgment, this covenant:

I will respect the hard-won scientific gains of those physicians in whose steps I walk, and gladly share such knowledge as is mine with those who are to follow.

I will apply, for the benefit of the sick, all measures [that] are required, avoiding those twin traps of over treatment and therapeutic nihilism.

I will remember that there is art to medicine as well as science, and that warmth, sympathy, and understanding may outweigh the surgeon's knife or the chemist's drug.

I will not be ashamed to say "I know not," nor will I fail to call in my colleagues when the skills of another are needed for a patient's recovery.

I will respect the privacy of my patients, for their problems are not disclosed to me that the world may know. Most especially must I tread with care in matters of life and death.

If it is given me to save a life, all thanks. But it may also be within my power to take a life; this awesome responsibility must be faced with great humbleness and awareness of my own frailty. Above all, I must not play at God.

I will remember that I do not treat a fever chart, a cancerous growth, but a sick human being, whose illness may affect the person's family and economic stability. My responsibility includes these related problems, if I am to care adequately for the sick.

I will prevent disease whenever I can, for prevention is preferable to cure.

I will remember that I remain a member of society, with special obligations to all my fellow human beings, those sound of mind and body as well as the infirm.

If I do not violate this oath, may I enjoy life and art, respected while I live and remembered with affection thereafter. May I always act so as to preserve the finest traditions of my calling and may I long experience the joy of healing those who seek my help.

—Written in 1964 by Louis Lasagna, Academic Dean of the School of Medicine at Tufts University, and used in many medical schools today.

Your Signature: _____

NATIONAL SCIENCE TEACHERS ASSOCIATION

Answer Key

Chapter 4
Teacher Guide for Earning Your White Coats: Medical School Research

TABLE 1.5. HIPPOCRATIC OATH DISCUSSION QUESTIONS

| As you read through the Hippocratic oath, which statements or terms puzzled you? |

| What concerns you about the Hippocratic oath? (Cons) | What seems promising about the Hippocratic oath? (Pros) |

| What is your stance on the Hippocratic oath? |
| What suggestions might you recommend if the Hipocratic Oath were revised again? |

References

Harvard Project Zero. Visible thinking. *www.old-pz.gse.harvard.edu/vt/VisibleThinking_ html_files/VisibleThinking1.html*

National Center for Biotechnology Information. PubMed health: Diabetes. *www.ncbi.nlm.nih.gov/pubmedhealth/PMH0002194*

National Center for Biotechnology Information. PubMed health: Sickle cell anemia. *www.ncbi.nlm.nih.gov/pubmedhealth/PMH0001554*

Tyson, P. 2001. The Hippocratic oath today. *www.pbs.org/wgbh/nova/body/hippocratic-oath-today.html*

WebMD. Health and pregnancy: Your guide to a healthy pregnancy. *www.webmd.com/baby/guide/getting-pregnant*

WebMD. HIV and AIDS: Overview and facts. *www.webmd.com/hiv-aids/default.htm*

Chapter 5
Teacher Guide for What's Wrong With Me, Doc?
Analyzing Medical Records

Before the Lesson

- Determine who your celebrity patients will be:

  - Patient #1 will be *diabetic (type 2)*. Be sure to select an overweight female. This female can be of any race or ethnicity. In this edition, Patient #1 is Jane Smith.

  - Patient #2 will be *pregnant*. Be sure to select a female. This female can be of any race or ethnicity. In this edition, Patient #2 is Cindy Jones.

  - Patient #3 will have *HIV*. Be sure to select a male of any race or ethnicity. In this edition, Patient #3 is John Thomas.

  - Patient #4 will have *sickle cell anemia*. Be sure to select a male of African American or Latino descent. In this edition, Patient #4 is Robert Smith.

- Edit the information found on the medical records to coincide with the celebrities you selected as your patients. For example, you may want to edit names, birth dates, occupations, and so on. Keep in mind, information doesn't necessarily have to be factually correct, but students get a good laugh when data accurately alludes to the characters inferred by the names.

- You may print photographs of your patients and hang them up on your classroom wall. (Note: We used generic names for this book.)

During the Lesson

- Students should read through each patient's medical record and reach a consensus regarding what health condition he or she may have based on the background research they have previously performed. Once each

Chapter 5
Teacher Guide for What's Wrong With Me, Doc? Analyzing Medical Records

team has developed a hypothesis for all of their patients, they should record it on their patients' designated medical charts under the section "Medical Record." Medical charts for Patients #1, #2, #3 and #4 can be found in Tables 2.1, 2.2, 2.3, and 2.4 (pp. 54–57, teacher edition), respectively. Students should cite three key pieces of evidence on their pads.

After the Lesson

- Ensure that each group has completed the "Medical Records" section of the patients' medical charts.
- Have a class debate regarding the conditions of each patient.
- *Note:* Students *may* get passionate about their initial hypotheses.

> **Answer Key**

Chapter 5
Teacher Guide for What's Wrong With Me, Doc? Analyzing Medical Records

What's Wrong With Me, Doc? Analyzing Medical Records

Task Overview

Congratulations! After applying to several residency programs, you finally found a job at Vanguard Hospital. Vanguard Hospital requires that you collaborate with your colleagues when treating patients. Your first day at work is extremely busy! Four celebrity patients have been rushed into your hospital with very alarming symptoms. Using your knowledge from medical school regarding the risk factors and symptoms of sickle cell anemia, HIV, pregnancy, and diabetes, hypothesize the condition afflicting each patient.

Medical Records

On the following eight pages are your four patients' medical records. Included in these records are their symptoms, lifestyle habits, and family medical histories. As a team, analyze these medical records. Remember, these medical records contain confidential information and the Hippocratic oath signed in medical school states that you "will respect the privacy of [your] patients, for their problems are not disclosed to [you, so] that the world may know." Therefore, it remains essential that you discuss your patients' health solely with your team. When your team has developed a hypothesis for each patient, record it on his or her designated medical chart under the section "Medical Record." Medical charts for Patients #1, #2, #3 and #4 can be found in Tables 2.1, 2.2, 2.3, and 2.4 (pp. 20–23, student edition) respectively. Be sure to cite evidence as to why you believe each patient has the health condition you suggest. These medical charts will play a vital role in your final diagnosis, so be as detailed as possible.

Chapter 5
Teacher Guide for What's Wrong With Me, Doc? Analyzing Medical Records

Answer Key

VANGUARD HOSPITAL
Medical Record

Patient's Name: <u>Jane Smith</u>
Date of Birth: <u>3/18/1954</u> Sex: <u>F</u> Height: <u>5'6''</u> Weight: <u>200 lbs</u>

Why are you here? <u>Had a dizzy spell and fainted. My vision seems blurry.</u>

Current Medications (prescription and non-prescription, vitamins, home remedies, birth control pills, herbs):
<u>None!</u>

Personal Medical History (please indicate whether you have had any of the following medical problems):

- ___ Congenital Heart Disease
- ___ Myocardial Infarction (heart attack)
- _X_ Hypertension (high blood pressure)
- ___ Diabetes (trouble regulating blood sugar)
- ___ High cholesterol (fat in blood)
- ___ Stroke (clogged artery to brain)
- ___ Coagulation (bleeding/clotting) disorder
- ___ Depression/suicide attempt
- ___ Alcoholism
- ___ If you ever had a blood transfusion
- ___ Abnormal pap smear (at gynecologist)
- ___ Other problems (specify) _____
- ___ Cancer
- _X_ Thyroid problem

Women's Gynecological History:
# of pregnancies <u>1</u>   # of deliveries <u>0</u>   # of abortions <u>1</u>   # of miscarriages <u>0</u>
Do you have any concerns about your periods? <u>No</u>
Do you have any concerns about menopause? <u>No</u>

Family History: Indicate with a check mark which family members have had any of the following:

Condition	Mom	Dad	Sis	Bro	Other relative	Condition	Mom	Dad	Sis	Bro	Other relative
Alcoholism						Hearing problems					
Anemia						Heart Attack (coronary artery disease)		X			
Arthritis (joint problems)						Hypertension (high blood pressure)	X	X			
Asthma						High cholesterol		X			
Bleeding Problems						Kidney disease					
Cancer	X					Migraine headaches					
Depression						Osteoporosis (weak bones)					
Diabetes, Type 1 (childhood onset)						Stroke		X			
Diabetes, Type II (adult onset)		X	X			Thyroid disorders	X		X		
Eczema (itchy skin)						Tuberculosis					
Epilepsy (seizures)						Neural disorders					
Genetic diseases											
Glaucoma (vision problem)											

Answer Key

Chapter 5
Teacher Guide for What's Wrong With Me, Doc? Analyzing Medical Records

Social History:

Tobacco use: ~~Quit:~~ Date _____
(Never) circled
~~Current smoker:~~ packs/day ____ # of yrs ____

Alcohol Use
Do you drink? <u>yes</u>
If yes, #drinks/wk <u>1/wk</u>

Do you use any recreational drugs? <u>no</u>
Have you ever used needles? <u>no</u>
Do you exercise regularly? <u>no</u>

Sexuality:
Are you sexually active? <u>yes</u>   Current sex partner(s) is/are: (Male) Female
Birth control method: <u>I'm on birth control pills.</u>
Do you practice safe sex? <u>We are in a committed relationship, so not always.</u>
Have you ever had a sexually transmitted disease (STDs)? <u>no</u>   If yes, please list: _____
Are you interested in being screened for a sexually transmitted disease? <u>Since I'm here, yes.</u>

Current Symptoms:

Constitutional
___ Fevers/chills/sweats
_X_ Unexplained weight loss/gain
_X_ Fatigue/Weakness
_X_ Excessive thirst or urination
Eyes
_X_ Change of vision
Ears/Nose/Throat/Mouth
___ Difficult hearing/ringing in ears
___ Problems with teeth/gums
___ Allergies
Cardiovascular
___ Chest pain/discomfort
___ Leg pain with exercise
___ Palpitations
Chest (breast)
___ Lump or discharge
Respiratory
___ Cough/Wheeze
_X_ Difficulty breathing
Gastrointestinal (digestive)
___ Abdominal pain
___ Blood in bowl movement
___ Nausea/vomiting/diarrhea

Genitourinary
___ Nighttime urination
___ leaking urine
___ Unusual vaginal bleeding
___ Discharge: penis or vagina
___ Sexual function problems
Musculo-skeletal
___ Muscle/joint pain
Skin
___ Rash or mole change
Neurological
_X_ Headaches
_X_ Dizziness/light-headedness
_X_ Numbness
___ Memory loss
___ Loss of coordination
Psychiatric
___ Anxiety/stress
_X_ Problems with sleep
___ Depression
Blood/Lymphatic (immune)
___ Unexplained lumps
___ Easy bruising/bleeding
Other: _____

Socio-economic:
Occupation: <u>Actress/Host/Editor</u>
Education completed: <u>High School</u>
Marital Status: <u>Not married</u>
Children: <u>0</u>
Who lives at home with you: <u>My boyfriend</u>

Chapter 5
Teacher Guide for What's Wrong With Me, Doc? Analyzing Medical Records

Answer Key

VANGUARD HOSPITAL
Medical Record

Patient's Name: <u>Cindy Jones</u>
Date of Birth: <u>12/8/69</u>   Sex: <u>F</u>   Height: <u>5'2''</u>   Weight: <u>120 lbs</u>

Why are you here? <u>Been feeling nauseous.</u>

Current Medications (prescription and non-prescription, vitamins, home remedies, birth control pills, herbs):
<u>birth control pills</u>

Personal Medical History (please indicate whether you have had any of the following medical problems):

___Congenital Heart Disease   ___Depression/suicide attempt
___Myocardial Infarction (heart attack)   ___Alcoholism
___Hypertension (high blood pressure)   ___If you ever had a blood transfusion
___Diabetes (trouble regulating blood sugar)   _X_Abnormal pap smear (at gynecologist)
___High cholesterol (fat in blood)   ___Other problems (specify) _____
___Stroke (clogged artery to brain)   ___Cancer
___Coagulation (bleeding/clotting) disorder   ___Thyroid problem

Women's Gynecological History:
\# of pregnancies <u>3</u>   \# of deliveries <u>2</u>   \# of abortions <u>1</u>   \# of miscarriages <u>0</u>
Do you have any concerns about your periods? <u>Yes</u>
Do you have any concerns about menopause? <u>No</u>

Family History: Indicate with a check mark which family members have had any of the following:

Condition	Mom	Dad	Sis	Bro	Other relative	Condition	Mom	Dad	Sis	Bro	Other relative
Alcoholism						Hearing problems					
Anemia	X		X			Heart Attack (coronary artery disease)					
Arthritis (joint problems)						Hypertension (high blood pressure)					
Asthma						High cholesterol					
Bleeding Problems						Kidney disease					
Cancer	X					Migraine headaches	X		X		
Depression		X	X			Osteoporosis (weak bones)					
Diabetes, Type 1 (childhood onset)						Stroke					
Diabetes, Type II (adult onset)						Thyroid disorders					
Eczema (itchy skin)						Tuberculosis					
Epilepsy (seizures)						Neural disorders					
Genetic diseases											
Glaucoma (vision problem)											

NATIONAL SCIENCE TEACHERS ASSOCIATION

Answer Key

Chapter 5
Teacher Guide for What's Wrong With Me, Doc? Analyzing Medical Records

Social History:

Tobacco use: Quit: Date <u>2 years ago ☺</u>   Alcohol Use:
    Never   Do you drink? <u>yes</u>
    Current smoker: packs/day ____ # of yrs ____   If yes, #drinks/wk <u>3/wk</u>

Do you use any recreational drugs? <u>no</u>
Have you ever used needles? <u>no</u>
Do you exercise regularly? <u>yes</u>

Sexuality:
Are you sexually active? <u>yes</u>   Current sex partner(s) is/are: (Male) Female
Birth control method: <u>I'm on birth control pills.</u>
Do you practice safe sex? <u>Since I'm on the pill, I don't always use condoms.</u>
Have you ever had a sexually transmitted disease (STDs)? <u>yes</u>   If yes, please list: <u>H.P.V.</u>
Are you interested in being screened for a sexually transmitted disease? <u>yes</u>

Current Symptoms:

Constitutional
___ Fevers/chills/sweats
X Unexplained weight loss/gain
X Fatigue Weakness
X Excessive thirst or urination
Eyes
___ Change of vision
Ears/Nose/Throat/Mouth
___ Difficult hearing/ringing in ears
___ Problems with teeth/gums
___ Allergies
Cardiovascular
___ Chest pain/discomfort
___ Leg pain with exercise
___ Palpitations
Chest (breast)
___ Lump or discharge
Respiratory
___ Cough/Wheeze
X Difficulty breathing
Gastrointestinal (digestive)
___ Abdominal pain
___ Blood in bowl movement
___ Nausea/vomiting/diarrhea

Genitourinary
___ Nighttime urination
___ Leaking urine
___ Unusual vaginal bleeding
___ Discharge: penis or vagina
___ Sexual function problems
Muscular-skeletal
___ Muscle/joint pain
Skin
___ Rash or mole change
Neurological
X Headaches
X Dizziness/light-headedness
___ Numbness
___ Memory loss
___ Loss of coordination
Psychiatric
___ Anxiety/stress
___ Problems with sleep
___ Depression
Blood/Lymphatic (immune)
___ Unexplained lumps
___ Easy bruising/bleeding
Other: _____

Socio-economic:
Occupation: <u>Singer/Actress</u>
Education completed: <u>High School</u>
Marital Status: <u>Single</u>
Children: <u>2</u>
Who lives at home with you: <u>My two kids</u>

Chapter 5
Teacher Guide for What's Wrong With Me, Doc? Analyzing Medical Records

Answer Key

VANGUARD HOSPITAL
Medical Record

Patient's Name: <u>John Thomas</u>
Date of Birth: <u>3/1/1989</u>   Sex: <u>M</u>   Height: <u>6'2''</u>   Weight: <u>140 lbs</u>

Why are you here? <u>Strange rash, chills, and fever</u>

Current Medications (prescription and non-prescription, vitamins, home remedies, birth control pills, herbs):
<u>none</u>

Personal Medical History (please indicate whether you have had any of the following medical problems):
___Congenital Heart Disease   ___Depression/suicide attempt
___Myocardial Infarction (heart attack)   ___Alcoholism
___ Hypertension (high blood pressure)   ___If you ever had a blood transfusion
___Diabetes (trouble regulating blood sugar)   ___Abnormal pap smear (at gynecologist)
___High cholesterol (fat in blood)   ___Other problems (specify) _____
___Stroke (clogged artery to brain)   ___Cancer
___Coagulation (bleeding/clotting) disorder   ___ Thyroid problem

Women's Gynecological History:
# of pregnancies ____   # of deliveries ____   # of abortions ____   # of miscarriages ____
Do you have any concerns about your periods? ____
Do you have any concerns about menopause? ____

Family History: Indicate with a check mark which family members have had any of the following:

Condition	Mom	Dad	Sis	Bro	Other relative	Condition	Mom	Dad	Sis	Bro	Other relative
Alcoholism		X				Hearing problems					
Anemia						Heart Attack (coronary artery disease)					
Arthritis (joint problems)						Hypertension (high blood pressure)	X	X			
Asthma						High cholesterol					
Bleeding Problems						Kidney disease					
Cancer:	X					Migraine headaches	X			X	
Depression		X				Osteoporosis (weak bones)					
Diabetes, Type 1 (childhood onset)						Stroke					X
Diabetes, Type II (adult onset)						Thyroid disorders					
Eczema (itchy skin)						Tuberculosis					
Epilepsy (seizures)						Neural disorders					
Genetic diseases											
Glaucoma (vision problem)		X									

NATIONAL SCIENCE TEACHERS ASSOCIATION

Answer Key

Chapter 5
Teacher Guide for What's Wrong With Me, Doc? Analyzing Medical Records

Social History:

Tobacco use: Quit: Date _____   Alcohol Use:
    Never   Do you drink? <u>yes</u>
    Current smoker: <u>1</u> packs/day # of yrs <u>3 years</u>   If yes, #drinks/wk <u>3/wk</u>

Do you use any recreational drugs? <u>no</u>
Have you ever used needles? <u>no</u>
Do you exercise regularly? <u>yes</u>

Sexuality:
Are you sexually active? <u>yes</u>   Current sex partner(s) is/are: Male (**Female**)
Birth control method: <u>condoms</u>
Do you practice safe sex? <u>sometimes</u>
Have you ever had a sexually transmitted disease (STDs)? <u>yes</u>   If yes, please list: <u>gonorrhea</u>
Are you interested in being screened for a sexually transmitted disease? <u>yes</u>

Current Symptoms:

Constitional
X Fever/chills/sweats
X Unexplained weight loss/gain
X Fatigue Weakness
___ Excessive thirst or urination
Eyes
X Change of vision
Ears/Nose/Throat/Mouth
___ Difficult hearing/ringing in ears
___ Problems with teeth/gums
___ Allergies
Cardiovascular
___ Chest pain/discomfort
___ Leg pain with exercise
___ Palpitations
Chest (breast)
___ Lump or discharge
Respiratory
___ Cough/Wheeze
___ Difficulty breathing
Gastrointestinal (digestive)
___ Abdominal pain
___ Blood in bowl movement
X Nausea/vomiting/diarrhea

Genitourinary
___ Nighttime urination
___ leaking urine
___ Unusual vaginal bleeding
X Discharge: penis or vagina
___ Sexual function problems
Musculo-skeletal
___ Muscle/joint pain
Skin
X Rash or mole change
Neurological
___ Headaches
X Dizziness/light-headedness
___ Numbness
___ Memory loss
___ Loss of coordination
Psychiatric
___ Anxiety/stress
___ Problems with sleep
___ Depression
Blood/Lymphatic (immune)
X Unexplained lumps
___ Easy bruising/bleeding
Other: _____

Socio-economic:
Occupation: <u>Singer/Dancer</u>
Education completed: <u>High School</u>
Marital Status: <u>Single</u>
Children: <u>Don't think so</u>
Who lives at home with you: <u>My mom, sometimes</u>

Chapter 5
Teacher Guide for What's Wrong With Me, Doc? Analyzing Medical Records

Answer Key

VANGUARD HOSPITAL
Medical Record

Patient's Name: <u>Robert Smith</u>
Date of Birth: <u>10/14/82</u>   Sex: <u>M</u>   Height: <u>5'8''</u>   Weight: <u>155 lbs</u>

Why are you here?: <u>had some difficulty breathing, been weak after recent concerts</u>

Current Medications (prescription and non-prescription, vitamins, home remedies, birth control pills, herbs):
<u>none</u>

Personal Medical History (please indicate whether you have had any of the following medical problems):

___Congenital Heart Disease   ___Depression/suicide attempt
___Myocardial Infarction (heart attack)   ___Alcoholism
___Hypertension (high blood pressure)   ___If you ever had a blood transfusion
___Diabetes (trouble regulating blood sugar) ___Abnormal pap smear (at gynecologist)
___High cholesterol (fat in blood)   ___Other problems (specify) _____
___Stroke (clogged artery to brain)   ___Cancer
___Coagulation (bleeding/clotting) disorder ___Thyroid problem

Women's Gynecological History:
# of pregnancies ____   # of deliveries ____   # of abortions ____   # of miscarriages ____
Do you have any concerns about your periods? ____
Do you have any concerns about menopause? ____

Family History: Indicate with a check mark which family members have had any of the following:

Condition	Mom	Dad	Sis	Bro	Other relative	Condition	Mom	Dad	Sis	Bro	Other relative
Alcoholism						Hearing problems					X (grandpa)
Anemia	X		X			Heart Attack (coronary artery disease)					
Arthritis (joint problems)						Hypertension (high blood pressure)					
Asthma						High cholesterol					
Bleeding Problems	X		X	X		Kidney disease					
Cancer:						Migraine headaches					
Depression		X				Osteoporosis (weak bones)					
Diabetes, Type 1 (childhood onset)						Stroke					
Diabetes, Type II (adult onset)						Thyroid disorders					
Eczema (itchy skin)						Tuberculosis					
Epilepsy (seizures)						Neural disorders					
Genetic diseases											
Glaucoma (vision problem)		X									

NATIONAL SCIENCE TEACHERS ASSOCIATION

Answer Key

Chapter 5
Teacher Guide for What's Wrong With Me, Doc? Analyzing Medical Records

Social History:

Tobacco use: Quit: Date <u>two months ago</u>   Alcohol Use:
    Never   Do you drink? <u>yes</u>
    Current smoker: _____ packs/day # of yrs _____   If yes, #drinks/wk <u>5/wk</u>

Do you use any recreational drugs? <u>Used to</u>
Have you ever used needles? <u>no</u>
Do you exercise regularly? <u>yes</u>

Sexuality:
Are you sexually active? <u>yes</u>   Current sex partner(s) is/are: Male (Female)
Birth control method: <u>condoms</u>
Do you practice safe sex? <u>every so often</u>
Have you ever had a sexually transmitted disease (STDs)? <u>yes</u>   If yes, please list: <u>Chlamydia</u>
Are you interested in being screened for a sexually transmitted disease? <u>yes</u>

Current Symptoms:

Constitutional
__ Fevers/chills/sweats
__ Unexplained weight loss/gain
X Fatigue Weakness
__ Excessive thirst or urination
Eyes
___ Change of vision
Ears/Nose/Throat/Mouth
___ Difficult hearing/ringing in ears
___ Problems with teeth/gums
___ Allergies
Cardiovascular
X Chest pain/discomfort
X Leg pain with exercise
___ Palpitations
Chest (breast)
___ Lump or discharge
Respiratory
___ Cough/Wheeze
X Difficulty breathing
Gastrointestinal (digestive)
X Abdominal pain
___ Blood in bowl movement
___ Nausea/vomiting/diarrhea

Genitourinary
___ Nighttime urination
___ Leaking urine
___ Unusual vaginal bleeding
___ Discharge: penis or vagina
___ Sexual function problems
Musculo-skeletal
X Muscle/joint pain
Skin
___ Rash or mole change
Neurological
___ Headaches
___ Dizziness/light-headedness
___ Numbness
___ Memory loss
___ Loss of coordination
Psychiatric
___ Anxiety/stress
___ Problems with sleep
___ Depression
Blood/Lymphatic (immune)
___ Unexplained lumps
___ Easy bruising/bleeding
Other: <u>My eyes and skin seem to have a yellowish color</u>

Socio-economic:
Occupation: <u>Rapper</u>
Education completed: <u>G.E.D.</u>
Marital Status: <u>Single</u>
Children: <u>??????</u>
Who lives at home with you: <u>Self</u>

Chapter 5
Teacher Guide for What's Wrong With Me, Doc? Analyzing Medical Records

TABLE 2.1. PATIENT #1'S MEDICAL CHART

Vanguard Hospital Medical Chart

Dr. _____   Patient #1: __Jane Smith__

Medical Records
I think they *might* be suffering from _____ because...
- Family history of Type 2 diabetes
- Symptoms: blurry vision, fatigue, weight loss, frequent urination/thirst

Lab 1—Urinalysis
☐ Normal ☑ Sugar ☐ Protein ☐ Bacteria

Lab 2—Digestive By-Products and BMI Analysis
☐ No issues ☑ Thirst ☐ Vomit ☐ Diarrhea ☐ Constipation
☐ BMI. # is __32__ ☐ Healthy ☐ Underweight ☐ Overweight ☑ Obese

Lab 3—Blood Smears
☑ Normal Red Blood Cells ☐ Sickle Cell
☐ # of Red Blood Cells: _____ million/µL/cu mm
☐ RBC count normal ☐ RBC too high ☐ RBC too low
☑ # of White Blood Cells: __7004__ /µL/cu mm
☑ WBC count normal ☐ WBC too high ☐ WBC too low

Lab 4—HIV Test
☐ + test   ☑ – test

Lab 5—Lung Capacity
Lung Volume #: __Approximately 2,000 cc__
☐ Capacity is normal ☐ Capacity is too high ☑ Capacity is too low

Lab 6—Hormone Test
☐ + for hCG o – for hCG ☐ No test necessary
☑ Sugar levels above 200 (low insulin) ☐ Sugar levels between 145 and 200 (borderline)
☐ Sugar levels below 145 (normal insulin)

Diagnosis: __Diabetes__

Treatments *(research this!)*:
- Insulin
- Exercise and diet
- Only cure would be a pancreas transplant

Prognosis *(research this!)*:
- Can live a normal, healthy life so long as she keeps disease under control.

TABLE 2.2. PATIENT #2'S MEDICAL CHART

Vanguard Hospital Medical Chart

Dr. _____   Patient #2: __Cindy Jones__

Medical Records
I think they *might* be suffering from _____ because...
- Symptoms: nauseous, weight gain, frequent urination, fatigue, difficulty breathing
- Lifestyle: on birth control pills but not using condoms regularly

Lab 1—Urinalysis
☑ Normal   ☐ Sugar   ☐ Protein   ☐ Bacteria

Lab 2—Digestive By-Products and BMI Analysis
☐ No issues   ☐ Thirst   ☐ Vomit   ☐ Diarrhea   ☐ Constipation
☐ BMI # is __22__   ☑ Healthy   ☐ Underweight   ☐ Overweight   ☐ Obese

Lab 3—Blood Smears
☑ Normal Red Blood Cells   ☐ Sickle Cell
☐ # of Red Blood Cells: _____ million/µL/cu mm
☐ RBC count normal   ☐ RBC too high   ☐ RBC too low
☑ # of White Blood Cells: __11,300__ /µL/cu mm
☐ WBC count normal   ☑ WBC too high   ☐ WBC too low

Lab 4—HIV Test
☐ + test   ☑ – test

Lab 5—Lung Capacity
Lung Volume #: __Approximately 1,000 cc__
☐ Capacity is normal   ☐ Capacity is too high   ☑ Capacity is too low

Lab 6—Hormone Test
☑ + for hCG   o – for hCG   ☐ No test necessary
☐ Sugar levels above 200 (low insulin)   ☑ Sugar levels between 145 and 200 (borderline)
☐ Sugar levels below 145 (normal insulin)

Diagnosis: __Pregnancy__

Treatments *(research this!)*:
- Prenatal vitamins
- Regular exercise

Prognosis *(research this!)*:
- Will deliver healthy baby if pregnancy is managed properly.

Chapter 5
Teacher Guide for What's Wrong With Me, Doc? Analyzing Medical Records

TABLE 2.3. PATIENT #3'S MEDICAL CHART

Vanguard Hospital Medical Chart

Dr. _____ Patient #3: __John Thomas__

Medical Records
I think they *might* be suffering from _____ because...
- Symptoms: rash, chills, fever, weight loss, discharge, unexplained lumps
- Lifestyle: practiced safe sex sometimes, has had sexually transmitted infection before (gonorrhea)

Lab 1—Urinalysis
☐ Normal ☐ Sugar ☒ Protein ☒ Bacteria

Lab 2—Digestive By-Products and BMI Analysis
☐ No issues ☐ Thirst ☐ Vomit ☒ Diarrhea ☐ Constipation
☐ BMI # is __18__ ☐ Healthy ☒ Underweight ☐ Overweight ☐ Obese

Lab 3—Blood Smears
☒ Normal Red Blood Cells ☐ Sickle Cell
☐ # of Red Blood Cells: _____ million/μL/cu mm
☐ RBC count normal ☐ RBC too high ☒ RBC too low
☒ # of White Blood Cells: __2,029__ /μL/cu mm
☐ WBC count normal ☐ WBC too high ☒ WBC too low

Lab 4—HIV Test
☒ + test ☐ − test

Lab 5—Lung Capacity
Lung Volume #: __Approximately 3,000 cc__
☐ Capacity is normal ☐ Capacity is too high ☒ Capacity is too low

Lab 6—Hormone Test
☐ + for hCG o − for hCG ☐ No test necessary
☐ Sugar levels above 200 (low insulin) ☐ Sugar levels between 145 and 200 (borderline)
☒ Sugar levels below 145 (normal insulin)

Diagnosis: __HIV__

Treatments *(research this!)*:
- Combination of antiretroviral drugs (ARVs)
- [Students may list specific drugs]
-

Prognosis *(research this!)*:
- [Students can determine patients prognosis based upon treatment.]

Answer Key

Chapter 5
Teacher Guide for What's Wrong With Me, Doc? Analyzing Medical Records

TABLE 2.4. PATIENT #4'S MEDICAL CHART

Vanguard Hospital Medical Chart

Dr. _____ Patient #4: ___Robert Smith___

Medical Records
I think they *might* be suffering from _____ because...
- Symptoms: difficulty breathing, fealing weak
- Family history: several blood disorders (anemia, bleeding)
- Symptoms (cont.): chest, leg, abdominal, joint pain; yellow tint to skin

Lab 1—Urinalysis
☑ Normal ☐ Sugar ☐ Protein ☐ Bacteria

Lab 2— Digestive By-Products and BMI Analysis
☑ No issues ☐ Thirst ☐ Vomit ☐ Diarrhea ☐ Constipation
☐ BMI # is __23__ ☑ Healthy ☐ Underweight ☐ Overweight ☐ Obese

Lab 3—Blood Smears
☐ Normal Red Blood Cells ☑ Sickle Cell
☐ # of Red Blood Cells: _____ million/µL/cu mm
☐ RBC count normal ☐ RBC too high ☐ RBC too low
☑ # of White Blood Cells: __9,001__ /µL/cu mm
☑ WBC count normal ☐ WBC too high ☐ WBC too low

Lab 4—HIV Test
☐ + test ☑ – test

Lab 5—Lung Capacity
Lung Volume #: __Approximately 2,500 cc__
☐ Capacity is normal ☐ Capacity is too high ☑ Capacity is too low

Lab 6—Hormone Test
☐ + for hCG o – for hCG ☐ No test necessary
☐ Sugar levels above 200 (low insulin) ☐ Sugar levels between 145 and 200 (borderline)
☑ Sugar levels below 145 (normal insulin)

Diagnosis: __sickle cell anemia__

Treatments *(research this!)*:
- Antibiotics (penicillin)
- Blood transfusions

Prognosis *(research this!)*:
Bouts of chronic pain; infection and organ failure could lead to death.

Chapter 6
Teacher Guide for Let's Diagnose Them, Lab 1
Urinalysis

Before the Lesson

- Review the excretory system with your students. Be sure to discuss:
  - the term *homeostasis*
  - the role of the excretory system, specifically the urinary track
  - the following organs: skin, lungs, the kidney, ureter, bladder, and urethra
  - problems that may arise in the urinary track: kidney stones, proteinuria, gout, urinary track infections, kidney failure
  - the medical terms: dialysis, urinalysis
- Obtain and set up the following supplies for the Urinalysis lab:

For the entire class:

- Patient #1 urine in a 250 ml beaker: Apple juice
- Patient #2 urine in a 250 ml beaker: Water with yellow food coloring
- Patient #3 urine in a 250 ml beaker: Egg Beaters diluted with water
- Patient #4 urine in a 250 ml beaker: Water with yellow food coloring
- 4 droppers, one for each beaker

For each group:

- 8 test tubes (2 labeled *Patient #1,* 2 labeled *Patient #2,* 2 labeled *Patient #3,* and 2 labeled *Patient #4)*
- 1 test tube rack
- 1 test tube clamp

Chapter 6
Teacher Guide for Let's Diagnose Them, Lab 1: Urinalysis

- 10 ml graduated cylinder
- 1 cup containing 20 ml of Benedict's solution
- 1 cup containing 20 ml of biuret reagent
- 2 droppers *(one for Benedict's, one for biuret)*
- 1 hot plate
- A 250 ml beaker of water
- 1 set of goggles per student
- Students should get *two* 6 ml samples of urine from all 4 patients

During the Lesson

- Establish roles for each student:
  - Task Manager—reads procedure and ensures everyone is following proper lab procedures
  - Materials Manager—retrieves and returns materials; cleans materials and table
  - Doctor(s)—completes the lab work (i.e., adding chemicals, heating chemicals, and so on); this role is to be completed by more than one group member
  - Recorder—ensures the group's data is recorded
- Review the procedure before allowing students to begin. Ask for any clarifying questions.
- Monitor student work. *Note:* Make sure students do NOT heat the urine samples during the protein indicator biuret test. Students often make the mistake of doing so.

After the Lesson

- Discuss urinalysis results with the class.
- Ensure that students have returned to their medical charts (Tables 2.1, 2.2, 2.3, and 2.4 on pages 20–23 in Chapter 2, student edition) and completed the section labeled "Lab 1— Urinalysis" for each of the four patients. Make sure they have checked off the evidence collected from each patient and have considered whether or not their original hypothesis is still supported or refuted by the evidence.

Answer Key

Chapter 6
Teacher Guide for Let's Diagnose Them, Lab 1: Urinalysis

Let's Diagnose Them, Lab 1: Urinalysis

In Medical School, you learned about the excretory system and its urinary track. Following proper procedure, you have asked each of your patients to submit urine samples. Nurses have informed you that both *Patient #1* and *Patient #2* are experiencing an abnormal increase in urination. Thankfully, a urinalysis can determine the state of one's health by examining physical and chemical properties of urine. Use the next few minutes to review your Medical School notes (Table 3.1) regarding the interpretation of urinalysis results. Once completed, examine your patients' urine samples.

TABLE 3.1. MEDICAL SCHOOL NOTES REGARDING URINALYSIS RESULTS

Color	If the urine is...	What it could indicate is ...
	Dark yellow	• dehydration or fever
	Pale light yellow	• patient drank a lot of liquids prior • diabetes
	Red with blood	• damage to kidneys
Odor	Fruity	• the presence of ketones (breakdown of fat), which is a product of diabetes or starvation
	Foul	• the presence of bacteria
Transparency	Clear	• normal urine samples appear clear/transparent
	Cloudy	• old samples could appear cloudy if bacteria has had time to grow on it • fresh samples could appear cloudy if a urinary track infection (UTI) is present (bacteria in the urethra) • fresh samples could appear cloudy if there are blood cells or pus
Sugar	Present	• patient ate a meal rich in carbohydrates prior to visit • a period of stress • diabetes
Protein	Present	• an abnormal condition called protein urea, that results from damage to kidneys

Materials

- 8 test tubes (2 each labeled *Patient #1*, *Patient #2*, *Patient #3*, and *Patient #4*)
- 1 test tube rack
- 1 test tube clamp

Chapter 6
Teacher Guide for Let's Diagnose Them, Lab 1: Urinalysis

Answer Key

- 10 ml graduated cylinder
- 1 cup containing 20 ml of Benedict's solution
- 1 cup containing 20 ml of biuret reagent
- 2 droppers (one for Benedict's, one for biuret)
- 1 hot plate
- 1 250 ml beaker of water
- 1 set of goggles for EACH member of the team
- 6 ml urine samples from all four patients (retrieve from your teacher)

Procedure

Setup

1. Label two test tubes *Patient #1*, two test tubes *Patient #2*, two test tubes *Patient #3*, and two test tubes *Patient #4*.

2. Place 6 ml of each patient's urine sample into his or her designated test tubes.

3. Add 150 ml of water to your 250 ml beaker and preheat it on a hot plate (*needed later for the sugar test*).

Physical Observations

1. Observe and describe the color, odor, and transparency of the four urine samples. Record physical descriptions in Table 3.2.

2. Get approval from teacher before advancing to Chemical Observations.

Chemical Observations

1. To test for sugar content, add approximately 6 ml of the TURQUOISE BLUE Benedict's solution to *one* test tube of *each* of the four patients' urine samples.

2. Place each patient's urine sample with the Benedict's solution into your hot water bath. Let sit for approximately five minutes.

3. If the TURQUOISE BLUE turns to ORANGE, as indicated by Table 3.3, sugar is present in the urine. Record whether or not sugar is present in Table 3.2.

4. To test for protein content, add approximately 2 ml of the BRIGHT BLUE biuret reagent into remaining test tubes for each of the urine samples. YOU DO NOT HEAT THESE TEST TUBES.

NATIONAL SCIENCE TEACHERS ASSOCIATION

Answer Key

Chapter 6
Teacher Guide for Let's Diagnose Them, Lab 1: Urinalysis

5. If the BRIGHT BLUE turns to a VIOLET PURPLE color, as indicated by Table 3.3, protein is present in the urine. Record whether or not protein is present in Table 3.2.

TABLE 3.2. RESULTS OF EACH PATIENT'S URINALYSIS

	Patient #1	Patient #2	Patient #3	Patient #4
Urination Habits	Excessive Urination	Excessive Urination	N/A	N/A
Color • Dark yellow? • Pale yellow? • Red with blood?	Pale yellow/ brown tint	Pale yellow	Bright yellow	Pale yellow
Odor • Fruity? • Foul? • Normal?	Fruity	Normal	Foul	Normal
Transparency • Cloudy? • Clear?	Clear	Clear	Cloudy	Clear
Is sugar present? • Yes: Turned orange (with heat) • No: Did <u>not</u> turn orange (with heat)	Yes	No	No	No
Is protein present? • Yes: Turned purple (with no heat) • No: Did <u>not</u> turn purple (with no heat)	No	No	Yes	No

TABLE 3.3. MEANING OF NUTRIENT INDICATOR TEST RESULTS

Nutrient	If nutrient is not present, color will remain ...	If nutrient is present, color will turn ...
Glucose (sugar)	Turquoise Blue	(with heat) Orange
Protein	Bright Blue	(no heat) Violet Purple

Chapter 6

Teacher Guide for Let's Diagnose Them, Lab 1: Urinalysis

Answer Key

Recall Questions

1. What is the role of the excretory system?

   ```
   The excretory system excretes liquid waste and monitors
   levels of salt, urea, and water in the body.
   ```

2. What is the function of the kidneys?

   ```
   The kidneys filter blood and remove toxins to be excreted by
   the urinary tract.
   ```

3. Why should a patient avoid eating a large meal before a urinalysis?

   ```
   Eating before a urinalysis can result in sugar in the
   urine. Therefore, a doctor might falsely diagnose his or
   her patient with diabetes.
   ```

4. Why should a patient provide a fresh sample of urine opposed to a sample that has sat out for several days?

   ```
   If urine samples are not fresh, bacteria can grow on the
   urine. Therefore, a doctor might falsely diagnose his or
   her patient with a urinary tract infection.
   ```

Critical Thinking Question

1. Dialysis is a medical device used to filter a patient's blood, when his or her kidneys can no longer function effectively. Dialysis is an expensive treatment and can cost as much as $500 per treatment. Medicare, the United States' medical insurance company for citizens over the age of 65, and for those with certain disabilities or kidney failure, covers most of the costs. However, under federal law, states are required to give emergency medical care to illegal immigrants, some of whom may require dialysis. As a result, taxpayers end up covering the cost. Considering your knowledge of the Hippocratic oath and the role of the excretory system, determine whether you would support or oppose this federal law. Justify your position.

   ```
   Answers will vary.
   ```

Conclusion

1. Look back at your medical notes in Table 3.1, and your lab results in Table 3.2. What could these results indicate about your patients?

   ```
   Patient #1's urine sample could indicate diabetes due to
   the pale urine sample, fruity odor, and the presence of
   ```

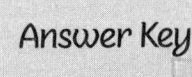

Answer Key

Chapter 6
Teacher Guide for Let's Diagnose Them, Lab 1: Urinalysis

```
sugar. Patient #3's urine sample could indicate a bacterial
infection, in addition to kidney failure, due to its cloudy
appearance, foul odor, and the presence of protein. Based
on this urinalysis, Patients #2 and #4 are in the clear.
```

2. Return to your patients' medical charts (Tables 2.1, 2.2, 2.3, and 2.4, pp. 20–23 in Chapter 2, student edition) and complete the section labeled "Lab 1—Urinalysis" for each of the four patients. Check off evidence collected from each patient and consider whether or not your original hypothesis is still supported or refuted by evidence.

References

Brumbak, K. Ga. patients get sicker as dialysis debate goes on. Associated Press, September 7, 2011. *www.romenewstribune.com/view/full_story/15422578/ article-Ga-patients-get-sicker-as-dialysis-debate-goes-on-?instance=home_ news_lead_ story.*

CostHelper. How much does dialysis cost? *http://health.costhelper.com/dialysis.html*

Mayo Clinic. Urinalysis: Results. *www.mayoclinic.com/health/urinalysis/MY00488/ DSECTION=results.*

Social Security Administration. Medicare benefits. *www.ssa.gov/pgm/medicare.htm*

Sullivan, J. 2004. Urinalysis. Columbia University Summer Research Program. *www. scienceteacherprogram.org/biology/Sullivan04.html*

Chapter 7

Teacher Guide for Let's Diagnose Them, Lab 2

Digestive By-Products and Body Mass Index Analysis

Before the Lesson

- Review nutrition and the digestive system with your students. Be sure to discuss:
    - the nutrition terms *organic molecule, protein, lipid, simple and complex carbohydrates, starch, calories, fiber, vitamins and minerals, portions, serving size, body mass index, glycemic index, metabolism*
    - the new federal guidelines for eating healthy in schools
    - the role of the digestive system
    - problems that may arise from poor nutrition: obesity, high blood pressure, heart attack, strokes, type 1 and 2 diabetes
    - the following organs: mouth, salivary glands, teeth, esophagus, epiglottis, stomach, small intestine, liver, pancreas, gallbladder, large intestine, colon, anus
    - the digestive terms: *mechanical digestion; chemical digestion;* and the enzymes *amylase, pepsin, HCl, bile, insulin*
    - problems that may arise in the digestive track: appendicitis, choking, constipation, diarrhea, vomiting, ulcers
- Visit our website (*www.StylishSchooling.com*) to download suggested activities and worksheets prior to starting this lab.
- Watch *Super Size Me*.
- Practice reading and analyzing various nutrient labels.
- Practice interpreting BMI charts[1]
- Compare and contrast the glycemic index values for different food items

---

1. The BMI chart included in this book is for adults.

Chapter 7
Teacher Guide for Let's Diagnose Them, Lab 2: Digestive By-Products and Body Mass Index Analysis

- Construct a meal plan template.
- Obtain and set up the following supplies for the Digestive By-product and BMI lab:

For the class:

- Patient #2 vomit in a 250 ml beaker: apple juice, grapes, and bananas
- Patient #3 diarrhea in a 250 ml beaker: water, starch, cottage cheese
- 2 spoons, one for each beaker

For each group of four:

- 6 test tubes (3 labeled Patient #2 and 3 labeled Patient #3)
- 1 test tube rack
- 1 test tube clamp
- 10 ml graduated cylinder
- 1 cup containing 20 ml of Benedict's solution
- 1 cup containing 20 ml of biuret reagent
- 1 cup containing 20 ml of Lugol's solution (iodine)
- 3 droppers (one for Benedict, one for biuret, one for Lugol's)
- 1 hot plate
- 250 ml beaker of water
- 1 set of goggles for each student
- 3 6ml samples of by-products from Patients #2 and #3

During the Lesson

- Establish roles for each student:
    - Task Manager—reads procedure and ensures that everyone is following proper lab procedures
    - Materials Manager—retrieves and returns materials; cleans materials and table

- Doctor(s)—completes the lab work (i.e., adding chemicals, heating chemicals, and so on); role may be completed by more than one group member
- Recorder—ensures that the group's data is recorded
- Review procedures before allowing students to begin. Ask for any clarifying questions.
- Monitor student work. *Note:* Again, make sure students do NOT heat the protein biuret test nor the starch Lugol test. Students often make the mistake of doing so.

After the Lesson

- Discuss digestive by-products and BMI analysis results with the class.
- Ensure that students have returned to their patients' medical charts (Tables 2.1, 2.2, 2.3, and 2.4 on pages 20–23 in Chapter 2, student edition) and completed the section labeled "Lab 2—Digestive By-Products and BMI Analysis" for each of the four patients. Make sure they have checked off the evidence collected from each patient and have considered whether or not their original hypothesis is still supported or refuted by the evidence.
- This is probably a good time to implement a small assessment on the excretory system, nutrition, and the digestive system to gauge where your students are in terms of understanding.

Chapter 7

Teacher Guide for Let's Diagnose Them, Lab 2: Digestive By-Products and Body Mass Index Analysis

Answer Key

Let's Diagnose Them, Lab 2: Digestive By-Products and Body Mass Index Analysis

Your patients have already spent one night at Vanguard Hospital and nurses on the nightshift had a very busy evening. The nurses just informed you that Patient #1 complained of excessive thirst, requesting water nearly every hour or so and Patient #2 was vomiting all morning and experiencing constipation. Additionally, Patient #3 experienced chronic episodes of diarrhea. The only patient with no major digestive issues was Patient #4, however he did complain of abdominal pain. Nurses are concerned that Patients #2 and #3 are losing vital nutrients such as glucose, starch, protein, lipids, vitamins, minerals, and water. It will be essential to replace whatever nutrients are missing with intravenous fluids.

Today your team of doctors will run tests on the digestive by-products of Patient #2 (vomit) and Patient #3 (diarrhea). Since there was no by-product from Patient #1 or #4 you will not be performing any tests for them. However, you will be expected to formulate an even deeper hypothesis about the conditions of Patient #1 and #4 based on their symptoms.

Lab Roles (Fill in Names of Team Members)

1. _____ is the task manager (reads procedure and ensures everyone is following proper protocol).

2. _____ is the materials manager (retrieves and returns materials; cleans materials and table).

3. _____ are the doctors (completes lab work, such as adding chemicals, heating chemicals, and so on; to be completed by more than one group member).

4. _____ is the recorder (ensures the group's data is properly recorded).

Materials

- 6 test tubes (3 labeled Patient #2 and 3 labeled Patient #3)
- 1 test tube rack
- 1 test tube clamp
- 10 ml graduated cylinder

Answer Key

Chapter 7
Teacher Guide for Let's Diagnose Them, Lab 2: Digestive By-Products and Body Mass Index Analysis

- 1 cup containing 20 ml of Benedict's solution
- 1 cup containing 20 ml of biuret reagent
- 1 cup containing 20 ml of Lugol's solution (iodine)
- 3 droppers (one for Benedict's, one for biuret, one for Lugol's)
- 1 hot plate
- 250 ml beaker of water
- 1 set of goggles (for EACH team member)
- 6 ml samples of digestive by-products from Patients #2 and #3 (retrieve from your teacher)

Procedure

Setup

1. Label three test tubes *Patient #2* and three test tubes *Patient #3*.
2. Place 6 ml of each patient's by-product into his or her designated test tubes.
3. Add 150 ml of water to your 250 ml beaker and preheat it on a hot plate *(needed later for the sugar test)*.

Physical Observations

1. Observe and describe the color and texture of the two patients' digestive by-products.
2. Record physical observations in Table 4.1.
3. Get approval from teacher before advancing to Chemical Observations.

Chemical Observations

1. To test for sugar content, add 6 ml of the TURQUOISE BLUE Benedict's solution to one test tube for each of the patient's digestive by-products.
2. Place each patient's digestive by-product with the Benedict's solution into your hot water bath. Let sit for approximately five minutes.
3. If the TURQUOISE BLUE turns to ORANGE, as indicated by Table 4.2, sugar is present in the digestive by-product. Record whether or not sugar is present in Table 4.1.

Chapter 7
Teacher Guide for Let's Diagnose Them, Lab 2: Digestive By-Products and Body Mass Index Analysis

Answer Key

4. To test for protein content, add approximately 2 ml of the BRIGHT BLUE biuret reagent remaining in the test tubes for each of the patients' digestive by-product samples. YOU DO <u>NOT</u> HEAT THESE TEST TUBES.

5. If the BRIGHT BLUE turns to a VIOLET PURPLE color, as indicated by Table 4.2, protein is present in the digestive by-product. Record whether or not protein is present in Table 4.1.

6. To test for starch content, add approximately five drops of the AMBER BROWN Lugol's solution into the left over test tubes for each of the patients' digestive by-product. Mix solution. YOU DO <u>NOT</u> HEAT THESE TEST TUBES.

7. If the AMBER BROWN turns to a PURPLE/BLACK color, as indicated by Table 4.2, starch is present in the digestive by-product. Record whether or not starch is present in Table 4.1.

TABLE 4.1. NUTRITION CONTENT FOUND IN PATIENT #2 AND #3'S DIGESTIVE BY-PRODUCTS

	Patient #1	Patient #2	Patient #3	Patient #4
Physical observations of by-products	Given: • thirsty all night • no vomit or diarrhea	vomit and constipation	diarrhea	Given: • abdominal pain • no digestive problems
Glucose • Present or not present?	N/A	Yes	No	N/A
Protein • Present or not Present?	N/A	No	Yes	N/A
Starch • Present or not Present?	N/A	No	Yes	N/A

NATIONAL SCIENCE TEACHERS ASSOCIATION

TABLE 4.2. MEANING OF NUTRIENT INDICATOR RESULTS

Nutrient	If nutrient is not present, color will remain…	If nutrient is present, color will turn…
Glucose (sugar)	Turquoise Blue	(with heat) Orange
Protein	Bright Blue	(no heat) Violet Purple
Starch	Amber/Brown	(no heat) Purple/Black

Nurses also mentioned they are concerned about the weight loss and gain the patients are experiencing. Some appear to have lost a significant amount of weight over a short period of time and some appear to be severely overweight or gaining weight at a rapid rate.

Body Mass Index (BMI)

1. Locate your patient's height (in feet and inches) and weight (in pounds) provided on each medical record (pages 12–19 in Chapter 2, student edition).

2. Using Figure 4.1, determine the BMI number for each of your patients.

3. Similarly, determine if your patients are underweight, overweight, obese, or healthy for their height using Table 4.3.

4. Record your findings in Table 4.4.

Chapter 7
Teacher Guide for Let's Diagnose Them, Lab 2: Digestive By-Products and Body Mass Index Analysis

Answer Key

FIGURE 4.1. BODY MASS INDEX (BMI) CHART FOR ADULTS

Weight in Pounds (lbs)

Height (ft.)	100	105	110	115	120	125	130	135	140	145	150	155	160	165	170	175	180	185	190	195	200
5'	19	20	21	22	23	24	25	26	27	28	29	30	31	32	33	34	35	36	37	38	39
5'1"	18	19	20	21	22	23	24	25	26	27	28	29	30	31	32	33	34	35	36	36	37
5'2"	18	19	20	21	22	22	23	24	25	26	27	28	29	30	31	32	33	33	34	35	36
5'3"	17	18	19	20	21	22	23	24	24	25	26	27	28	29	30	31	32	32	33	34	35
5'4"	17	18	18	19	20	21	22	23	24	25	25	26	27	28	28	29	30	31	32	33	33
5'5"	16	17	18	19	20	20	21	22	23	24	25	26	26	27	28	29	30	31	32	32	33
5'6"	16	17	17	18	19	20	21	21	22	23	24	25	25	26	27	28	29	29	30	31	32
5'7"	15	16	17	18	18	19	20	21	22	22	23	24	25	26	27	27	28	29	30	31	31
5'8"	15	16	16	17	18	19	20	21	22	22	23	24	25	26	27	27	28	29	29	30	31
5'9"	14	15	16	17	17	18	19	20	20	21	22	23	24	24	25	27	27	28	29	28	28
5'10"	14	15	15	16	17	18	18	19	20	21	21	22	23	24	24	25	26	27	28	28	28
5'11"	14	14	15	15	16	17	18	18	19	20	21	22	22	23	23	24	25	26	26	27	27
6'	13	14	14	15	16	17	17	18	19	20	20	21	22	22	23	24	25	25	26	27	27
6'1"	13	13	14	15	15	16	17	17	18	19	19	20	21	21	22	23	23	24	25	25	26
6'2"	12	13	14	14	15	16	16	17	18	18	19	19	20	21	21	22	23	23	24	25	25

TABLE 4.3. BMI RANGES

Underweight	Healthy	Overweight	Obese
<18.9	19–24.9	25–29.9	>30

TABLE 4.4. THE BMI VALUE FOR EACH PATIENT

	Patient #1	Patient #2	Patient #3	Patient #4
Height	5' 6"	5' 2"	6' 2"	5' 8"
Weight	200	120	140	155
BMI #	32	22	18	24
Are they ... Healthy? Underweight? Overweight? Obese?	Obese	Healthy	Underweight	Healthy

Recall Questions

1. What is the role of the digestive system?

   The digestive system breaks down food into its nutritional components, specifically so our cells can get energy.

2. What is the difference between simple carbohydrates and complex carbohydrates?

   Simple carbohydrates are easy for our body to break down. They provide short-term energy and often lead to quick bursts of energy (sugar spikes). Complex carbohydrates are harder for our body to break down. They provide long-term energy and lead to stabilized blood sugar levels.

3. What is the glycemic index (GI)?

   The glycemic index is a ranking system, on a scale of 0-100, which indicates how fast carbohydrates raise blood sugar levels after eating.

4. What foods are considered high on the glycemic index? What foods are considered low?

   Foods that are considered high on the glycemic index are simple carbohydrates such as white rice; white pasta; white bread; white potatoes; and other sugar-laden candies, pastries, and beverages. Foods that are considered low on the glycemic index are complex carbohydrates such as brown rice, whole-wheat pasta and breads, sweet potato, most fruits, vegetables, meats, and nuts.

5. Why is a high BMI value considered worrisome for doctors?

   People with a high BMI values are at greater risk of developing high blood pressure (hypertension), high LDL

Chapter 7
Teacher Guide for Let's Diagnose Them, Lab 2: Digestive By-Products and Body Mass Index Analysis

Answer Key

cholesterol (bad one), low HDL cholesterol (good one), high blood glucose levels (potentially type 2 diabetes), and may, potentially, lead to death.

6. Explain why BMI charts, although used by doctors, may not accurately depict one's true state of health.

   BMI is a good way to assess if someone is overweight or underweight for his or her height, age, and sometimes, gender, but it does not indicate how healthy a person is internally. Additionally, muscle mass weighs more than fat, and can result in a higher BMI value.

Critical Thinking Question

1. During summer 2012, in an attempt to curb childhood obesity, the U.S. Department of Food and Agriculture mandated schools to offer nutritious meals at breakfast and lunch. Some of the guidelines drafted include:

   - offering vegetables, fruit, whole-grains, meat, meat alternatives, and fat-free unflavored milk daily and at each meal time
   - reducing the sodium content of meals over a 10-year period
   - preparing meals that contain 0 grams of trans fats, and
   - designing meals that target the specific caloric needs of varying age groups

   Suppose you were selected to advise the Department of Food and Agriculture; using your knowledge of nutrition and its effects on a person's health, construct a five-day meal plan, for *either* breakfast or lunch, that would be considered acceptable under these new regulations. Be sure to include the nutrition information in the meal plan (amounts of sodium, trans fats, carbohydrates, and fiber; total calories; and so on).

   Answers may vary.

Conclusion

1. Look back at your lab results in Table 4.1 and Table 4.4. What could these results indicate about your patients?

   Patient #1 is experiencing excessive thirst and she is considered obese based on the BMI chart. Both of these are symptoms of type 2 diabetes. Patient #2 is experiencing vomit and constipation, which are symptoms of pregnancy. Her weight is still normal, so she could be in an early stage of pregnancy. Patient #3 is experiencing diarrhea,

which is more common in HIV patients than with any of
the other health conditions. He is also underweight. And
finally, Patient #4 is appears to have abdominal cramps. He
could be experiencing a pain crises. His weight, at the
moment, seems fine.

2. Return to your patients' medical charts (Tables 2.1, 2.2, 2.3, and 2.4, pp. 20–23 in Chapter 2, student edition) and complete the section labeled "Lab 2—Digestive By-Products and BMI Analysis" for each of the four patients. Check off evidence collected from each patient and consider whether or not your original hypothesis is still supported or refuted by evidence.

References

Brand-Miller, J. About glycemic index. University of Sydney. *www.glycemicindex.com/about.php*

Centers for Disease Control and Prevention. Body mass index. *www.cdc.gov/healthyweight/assessing/bmi*

Free freebmicalculator.net. BMI Chart—Your body mass index in English units. *www.freebmicalculator.net/bmi-chart.php*

Harvard Medical School. 2013. Glycemic index and glycemic load for 100+ foods. Harvard Medical Publications. *www.health.harvard.edu/newsweek/ Glycemic_index_ and_glycemic_load_for_100_foods.htm*

National Blood Heart and Lung Institute. Assessing your weight and health risk. *www.nhlbi.nih.gov/health/public/heart/obesity/lose_wt/risk.htm*

National Blood Heart and Lung Institute. Calculate your body mass index. *http://nhlbisupport.com/bmi*

U.S. Government Printing Office. 2012. Nutrition standards in the national school lunch and school breakfast programs. *Federal Register.* 77 (17). 4087–4167. *www.gpo.gov/fdsys/pkg/FR-2012-01-26/pdf/2012-1010.pdf*

Chapter 8
Teacher Guide for Let's Diagnose Them, Lab 3
Blood Smears

Before the Lesson

- Review the circulatory system with your students. Be sure to discuss:
    - the role of the circulatory system
    - the following blood vessels: *veins*, *arteries*, and *capillaries*
    - the following parts of the heart: *superior* and *inferior vena cava*, *right atria*, *right ventricle*, *pulmonary artery*, *pulmonary vein*, *left atria*, *left ventricle*, and *aorta*
    - the following parts of blood: *red blood cells*, *white blood cells*, *plasma*, and *platelets*.
    - problems that may arise from poor nutrition or genetics: *high blood pressure*, *heart attack*, *strokes*, *hemophilia*, and *sickle cell anemia*.
- Visit our website (*www.StylishSchooling.com*) to download suggested activities and worksheets prior to starting this lab:
    - If your class hasn't watched *Super Size Me* yet, this would also be a good time to show it
    - Have students memorize the *Vena Cava* song for extra credit
    - Dissect a sheep's heart
    - Review how to focus a compound light microscope
- Obtain and set up the following supplies for the Blood Smears lab.

For each group of four:

- 1–2 compound light microscopes

Chapter 8
Teacher Guide for Let's Diagnose Them, Lab 3: Blood Smears

- Provide 3 prelabeled, normal blood smears (Patient #1, Patient #2, and Patient #3)
- Provide 1 prelabeled sickle cell blood smear (Patient #4) per group

Note: Sometimes students have difficulty observing the slides under the microscope. As an alternative to the microscope, you could print out normal and sickle blood smears from the internet and label them Patient #1–#4.

During the Lesson

- Establish roles for each student:
    - Task Manager—reads procedure and ensures that everyone is following proper lab procedures
    - Materials Manager—retrieves and returns materials; cleans materials and table
    - Doctor(s)—completes the lab work (i.e., focusing the microscope); role is to be completed by more than one group member
    - Recorder—ensures that the group's data is recorded
- Review the procedure before allowing students to begin. Ask for any clarifying questions.
- Monitor student work. Ensure that students are making proper observations and sketching accordingly.

After the Lesson

- Discuss Blood Smear results with the class.
- Ensure that students have returned to their patients' medical charts (Tables 2.1, 2.2, 2.3, and 2.4 on pages 20–23 in Chapter 2, student edition) and completed the section labeled "Lab 3—Blood Smears" for each of the four patients. Make sure they have checked off the evidence collected from each patient and have considered whether or not their original hypothesis is still supported or refuted by the evidence.

Answer Key

Chapter 8
Teacher Guide for Let's Diagnose Them, Lab 3: Blood Smears

Let's Diagnose Them, Lab 3: Blood Smears

Some doctors on your team are beginning to think that some of your patients' symptoms may be caused by either a pathogen or a genetic disorder. A pathogen causes harm or disease in another living organism. Examples include viruses, bacteria, and fungi. Genetic disorders are diseases inherited from one's parents.

Today your team of doctors will analyze the red blood cells (RBCs) of patients under a microscope. Nurses have also provided you with your patients' red blood cell and white blood cell counts. Use your medical school notes (Table 5.1) as a reference for diagnosing your patients.

TABLE 5.1. MEDICAL SCHOOL NOTES REGARDING RED AND WHITE BLOOD CELLS

	Function	Healthy if ...	Unhealthy if ...
Red blood cells (RBC)	Uses the protein, hemoglobin, to carry oxygen around the body	Shaped like a donut Female RBC count = 4.2–5.4 million/µL/cu mm Male RBC count = 4.7–6.1 million/µL/cu mm	Shaped like a sickle, indicating a genetic disorder called **sickle cell anemia**. If *lower* than normal, could indicate anemia, such as **sickle cell anemia**. However, anemia is also common during the first six months of **pregnancy.** If *higher* than normal, could indicate polycythaemia, a disorder of the bone marrow.
White bblood cells (WBC)	Help fight infections by (A) Phagocytosis of foreign agents (B) Producing antibodies against foreign agents	WBC count = 4,300–10,800 cells/µL/cu mm	If *lower* than normal, could indicate viral infections like **HIV**, low immunity and bone marrow failure. If *higher* than normal, could indicate infection, systemic illness, inflammation, allergy, leukemia, and tissue injury caused by burns, or **pregnancy**.

Lab Roles (Fill in Names of Team Members)

1. _____ is the task manager (reads procedure and ensures everyone is following proper protocol).

2. _____ is the materials manager (retrieves and returns materials; cleans materials and table).

3. _____ are the doctors (completes lab work, such as adding chemicals, heating chemicals, and so on; to be completed by more than one group member).

4. _____ is the recorder (ensures the group's data is properly recorded).

Chapter 8
Teacher Guide for Let's Diagnose Them, Lab 3: Blood Smears

Answer Key

Materials

- 1–2 compound light microscopes
- Blood smears from your four patients, provided by your teacher

Procedure

1. Start with the microscope stage as far away from the lens as possible.
2. Place Patient #1's blood smear on the stage and secure it with the stage clips.
3. Place the objective lens to low power (4×).
4. Using the coarse adjustment (big knob), begin to focus the slide.
5. Once focused, change the objective lens to medium power (10×).
6. Using the coarse adjustment (big knob), begin to focus the slide.
7. Once focused, change the objective lens to high power (40×).
8. Using the fine adjustment (small knob), begin to focus the slide.
9. Sketch your observation of red blood cells at the power most easily observable in Table 5.2.
10. Repeat steps 1–9 for patients #2, #3, and #4.

Answer Key

Chapter 8
Teacher Guide for Let's Diagnose Them, Lab 3: Blood Smears

TABLE 5.2. THE BLOOD SMEAR RESULTS FOR EACH PATIENT

	Patient #1	Patient #2	Patient #3	Patient #4
Sketch a detailed picture of what you observe here:				
RBC Shape Normal or Sickle?	Normal	Normal	Normal	Sickle
# of RBCs million/µL/cu mm Normal, high, or low?	4.4 Normal	3.0 Low	5.1 Normal	3.2 Low
# of WBCs /µL/cu mm Normal, high, or low?	7,004 Normal	11,300 High	2,029 Low	9,001 Normal

Recall Questions

1. What is the role of the circulatory system?

   The circulatory system transports nutrients to, and waste away from, cells in the body.

2. How do arteries differ from veins?

   Arteries are blood vessels that carry oxygenated blood away from the heart toward cells. Veins are blood vessels that carry deoxygenated blood toward the heart away from cells.

3. What problems could arise in the circulatory system from poor nutrition and lack of exercise?

   Problems that can arise from poor nutrition and lack of exercise include heart attacks, strokes, blood clots, high blood pressure, etc.

4. What is the role of a red blood cell?

   Red blood cells carry oxygen molecules throughout the body.

Chapter 8
Teacher Guide for Let's Diagnose Them, Lab 3: Blood Smears

5. How does sickle cell anemia differ from sickle cell trait?

   ```
   Sickle cell anemia requires two copies of the sickle cell
   gene (one from mom, one from dad). A person with sickle
   cell anemia exhibits all the symptoms of the disease.
   Sickle cell trait requires one copy of the sickle cell gene
   (one from mom or dad). A person with sickle cell trait does
   not exhibit all the symptoms of the disease and is often
   malaria resistant.
   ```

6. What happens to the hemoglobin protein on a red blood cell if someone has sickle cell anemia?

   ```
   The hemoglobin protein is mutated in a person with sickle
   cell anemia. Rather than carrying the normal four oxygen
   molecules, a person with sickle cell can only carry half of
   the amount.
   ```

Critical Thinking Question

1. In early 2012, coaches instructed a Pittsburgh professional football player, with the sickle cell *trait*, to sit out a game in the high-altitude city, of Denver, Colorado. Doctors claimed that the trait, in combination with extreme physical activity and a high altitude, was the primary reason he needed to have his spleen and gallbladder removed after a previous game in the city. However, it has been estimated that at least 90 other NFL players carry the sickle cell trait, and of those who have played in Denver, they have never experienced such issues before. In fact, a study performed by Howard University in 2000, showed no complications in athletes carrying sickle cell trait during the Mexico City Olympics, another high-altitude location. Suppose you were a coach of a high school, college, or professional sports team. Knowing what you know about sickle cell anemia and the trait, how would you handle a situation similar to this one, in which one of your players has sickle cell trait or the disease? Justify your position.

   ```
   Answers may vary.
   ```

Conclusion

1. Look back at your medical notes in Table 5.1, and your lab results in Table 5.2. What could these results indicate about your patients?

   ```
   Patient #2 has normal red blood cells, but a low red blood
   cell count. Her white blood cell count, however, is high.
   ```

Answer Key

Chapter 8
Teacher Guide for Let's Diagnose Them, Lab 3: Blood Smears

> This could indicate that she is pregnant. Patient #3 has normal red blood cells and a normal red blood cell count. However, his white blood cell count is too low. This could indicate that he has HIV. Patient #4 has abnormally shaped red blood cells and a low red blood cell count. His white blood cell count is normal. He can officially be diagnosed as having sickle cell anemia based on his blood smear. Patient #1's red blood cells and white blood cells appear normal.

2. Return to your patients' medical charts (Tables 2.1, 2.2, 2.3, and 2.4, pp. 20–23 in Chapter 2, student edition) and complete the section labeled "Lab 3—Blood Smears" for each of the four patients. Check off evidence collected from each patient and consider whether or not your original hypothesis is still supported or refuted by evidence.

References

National Institutes of Health. MedLine Plus—RBC Count. *www.nlm.nih.gov/medlineplus/ency/article/003644.htm*

National Institutes of Health. MedLine Plus—WBC Count. *www.nlm.nih.gov/medlineplus/ency/article/003643.htm*

Strauss, C. 2012. Clark's sickle cell scare an anomaly. *The Daily*. January 5. *http://msn.foxsports.com/nfl/story/pittsburgh-steelers-ryan-clark-sickle-cell-trait-scare-an-anomaly-010512*

Chapter 9

Teacher Guide for Let's Diagnose Them, Lab 4

HIV Test

Before the Lesson

- Review the immune system with your students. Be sure to discuss:
    - the role of the immune system
    - the following organs: *spleen, thymus, bone marrow,* and *lymph nodes*
    - the following white blood cells and their functions: *T-cells* and *B-cells*
    - the difference between viral infections and bacterial infections
    - the terms: *vaccines, antibodies,* and *antigens*
    - problems that may arise in the immune system:
        - *viral sexually transmitted infections: HIV, HPV, herpes simplex I and II, hepatitis C;*
        - *bacterial sexually transmitted infections: chlamydia, gonorrhea, syphilis*
    - other infections or health conditions of student interest: *chicken pox, mononucleosis, measles, leukemia, autoimmune disorders*
- Visit our website (*www.StylishSchooling.com*) to download suggested activities/worksheets prior to starting this lab:
    - Watch the movie *Osmosis Jones* (best for less mature audiences) or *And the Band Played On* (best for more mature audiences)
    - Show statistics for sexually transmitted infections by age, race, region, and so on
    - "Dear Abby" writing prompts
- Obtain and set up the following supplies for the HIV lab:

Chapter 9
Teacher Guide for Let's Diagnose Them, Lab 4: HIV Test

For the entire class:

Part I: Round 1

- Class set of cups labeled *Control 1* and *Round 1* containing distilled water. These cups represent "no virus."
- 2 cups also labeled *Control 1* and *Round 1* containing diluted sodium hydroxide (NaOH). These cups represent "the virus."
- Phenolphthalein (the "HIV indicator")
- 1 dropper
- Distilled water
- 1 set of goggles per group member

Part I: Round 2

- Class set of cups labeled *Control 2* and *Round 2* containing distilled water. These cups represent "no virus."
- 2 cups also labeled "Control 2" and "Round 2" containing diluted sodium hydroxide (NaOH) . These cups represent "the virus."
- Phenolphthalein (the "HIV indicator")
- 1 dropper
- Distilled water

Part II

- Bodily fluid samples from all four patients. Patients #1, #2 and #4 consist of distilled water. Patient #3 consists of 5 ml diluted NaOH.

Part II

For each group of four:

- 4 test tubes (labeled *Patient #1*, *Patient #2*, *Patient #3*, and *Patient #4*, respectively)
- 1 cup with phenolphthalein (the "HIV indicator")
- 1 dropper
- Bodily fluid samples from all four patients

During the Lesson

- Establish roles for each student:
    - Task Manager—reads procedure and ensures that everyone is following proper lab procedures
    - Materials Manager—retrieves and returns materials; cleans materials and table
    - Doctor(s)—completes the lab work (i.e. add chemicals)
    - Recorder—ensures that the group's data is recorded
- Review the procedure before allowing students to begin. Ask for any clarifying questions.
- Part I: Give students approximately three minutes to find their five partners. The class might be a bit rowdy during this portion of the lab, so have a system in place for moving forward (e.g., light flickering, countdown method, a bell). Another suggestion would be to instruct students to sit back in their seats when they have already found five people.
- Part I (continued): It is also recommended that you have a clean set of cups ready for the second round with two partners.

After the Lesson

- Discuss HIV results with the class.
- Ensure that students return to their patients' medical charts (Tables 2.1, 2.2, 2.3, and 2.4 pages 20-23 in Chapter 2, student edition) and complete the section labeled "Lab 4—HIV Test" for each of the four patients. Make sure they have checked off the evidence collected from each patient and have considered whether or not their original hypothesis is still supported or refuted by the evidence.
- Now might be a good time to implement a second assessment on the circulatory system and the immune system.

Chapter 9
Teacher Guide for Let's Diagnose Them, Lab 4: HIV Test

Answer Key

Unit 6: Let's Diagnose Them, Lab 4: HIV test

Based on blood samples recently analyzed, one of your patients, Patient #3, had a very low white blood cell (WBC) count. This is very alarming to you and your team of doctors since it might indicate the patient is suffering from a viral infection. Since all of your patients have requested an STD test on their medical records, you will specifically check for antibodies produced against the Human Immunodeficiency Virus (HIV).

Before you test your patients, it is essential that you discuss the importance of safe sex. As you have already noticed, all patients indicated on their medical records that they are currently sexually active, yet none of them reported the regular use of condoms as a method of protection. In order to emphasize the importance of safe sex, you and the other doctors in the hospital will demonstrate a simulation of how fast an STD can travel within a population of multiple sex partners who are engaging in unsafe sex. Upon completing this "sex-talk"/demonstration, you will then test your patients.

Lab Roles (Fill in Names of Team Members)

1. _____ is the task manager (reads procedure and ensures everyone is following proper protocol).

2. _____ is the materials manager (retrieves and returns materials; cleans materials and table).

3. _____ are the doctors (completes lab work, such as adding chemicals, heating chemicals, and so on; to be completed by more than one group member).

4. _____ is the recorder (ensures the group's data is properly recorded).

Materials

Part I

- 4 cups per doctor: 1 filled with NaOH ("virus") and 3 with H_2O ("no virus") designated by the teacher
- Phenolphthalein (the "HIV indicator")
- 1 dropper
- Soap

Answer Key

Chapter 9
Teacher Guide for Let's Diagnose Them, Lab 4: HIV Test

- Distilled water
- 1 set of goggles

Part II

- 4 test tubes (labeled Patient #1, Patient #2, Patient #3, and Patient #4, respectively)
- 1 cup with phenolphthalein (the "HIV indicator")
- 1 dropper
- Bodily fluid samples from all four patients

Procedure

Part I: Done as a Class

ROUND 1: Multiple Sexual Partners

1. Obtain two cups labeled "Control 1" and "Round 1" from your teacher. The fluids inside the two cups are the same, and represent vaginal fluid and/or semen. Everyone in the class will have H_2O ("no virus") in their two cups except for *one* person who will have NaOH ("virus").

2. When your teacher says "Go!" find *five* different people to have "unprotected sex" with, represented by the exchange of fluids in your "Round 1" cup. Make sure your "Control 1" cup is off to the side and remains uncontaminated.

3. Return to your seat after changing fluids with five different people.

4. Upon completion, the head doctor (your teacher) will give each of you an HIV test. The head doctor will place 1 drop of phenolphthalein (the "HIV indicator") into your "Round 1" cups.

5. If your "Round 1" cup remains clear, you did *not* contract HIV. If the fluid turns pink, you did contract HIV. Record your observations in Table 6.1.

6. Who initiated the infection? The head doctor will place 1 drop of phenolphthalein (the "HIV indicator") into your "Control" cup. Record your observations in Table 6.1.

7. Dispose of your cups.

Chapter 9
Teacher Guide for Let's Diagnose Them, Lab 4: HIV Test

Answer Key

DATA TABLE 6.1. RESULTS FROM MULTIPLE SEX PARTNER DEMONSTRATION

Color of the fluid my "Round 1" cup after HIV test:	Answers will vary.
Total number of "doctors" with HIV after exchanging fluids with five different partners:	Answers will vary.
The person who initially transmitted the HIV infection to everyone in the room was:	Answers will vary.

ROUND 2: Limited Sexual Partners

1. Obtain two new cups (labeled "Control 2" and "Round 2") from your teacher. Again, the fluids inside the two cups are the same, and represent vaginal fluid and/or semen. Everyone in the class will have H_2O ("no virus") in their two cups except for *one* person who will have NaOH ("virus").

2. When your teacher says "Go!" find *two* different people to have "unprotected sex" with, represented by the exchange of fluids in your "Round 2" cups. Make sure your "Control 2" cup is off to the side and remains uncontaminated.

3. Return to your seat after exchanging fluids with two different people.

4. Upon completion, the head doctor (your teacher), will give each doctor an HIV test. The head doctor will place 1 drop of phenolphthalein (the "HIV indicator") into your "Round 2" cup.

5. If the fluid in your "Round 2" cup remains clear, you did *not* contract HIV. If the test tube turns pink, you did contract HIV. Record your observations in Table 6.2.

6. Who initiated the infection? The head doctor will place 1 drop of Phenolphthalein (the "HIV indicator") into your "Control" cup. Record your observations in Table 6.2.

7. Dispose of your cups.

TABLE 6.2. RESULTS FROM LIMITED SEX PARTNER DEMONSTRATION

Color of fluid in "Round 2" cup after test:	Answers will vary.
Total number of "doctors" with HIV after exchanging fluid with two partners?	Answers will vary (but the total number should be less than in the previous round)
The person who initially transmitted the HIV infection to everyone in the room was:	Answer will vary.

Answer Key

Chapter 9
Teacher Guide for Let's Diagnose Them, Lab 4: HIV Test

Part II: Done as a Team of Doctors

1. Label four test tubes *Patient #1, Patient #2, Patient #3,* and *Patient #4,* respectively.

2. Obtain bodily fluids from each patient and place the samples in their designated test tubes.

3. Perform the HIV test on each of them by adding 1 drop of phenolphthalein (the "HIV indicator") to their semen or vaginal fluid sample.

4. Record the color in Table 6.3 and determine whether or not they have HIV.

TABLE 6.3. RESULTS FROM EACH PATIENT'S HIV TEST

	Patient #1	Patient #2	Patient #3	Patient #4
Color • Clear or Pink?	Clear	Clear	Pink	Clear
HIV status • HIV + or –?	Negative	Negative	Positive	Negative

Recall Questions

1. What is the role of the immune system?

   The immune system fights infections and protects the body from foreign substances.

2. What is the role of a white blood cell?

   White blood cells identify and destroy pathogens that enter the body.

3. At what point does HIV become AIDS?

   HIV becomes AIDS when the white blood cell count (specifically, T-cells) drops below 200.

4. Why does someone with AIDS become more susceptible to other infections?

   A person with AIDS becomes more susceptible to other infections because his or her white blood cell count is very low. As a result, they are unable to fight off pathogens that enter the body.

Chapter 9
Teacher Guide for Let's Diagnose Them, Lab 4: HIV Test

Answer Key

5. What are some ways a person can prevent the spread of HIV to others?

   `Abstinence, using protection during sexual intercourse and interaction, avoiding blood-to-blood contact, and preventative measures between an infected mother and her child.`

6. List one difference between a bacterial infection and a viral infection.

   `A bacterial infection can be cured and bacteria are considered living. A viral infection cannot be cured, only treated. Viruses are not considered living, unless they are within a host.`

7. Approximately how long after the initial infection will HIV antibodies show up in an HIV test?

   `It can take up to three months after the initial infection for HIV antibodies to show up on an HIV test.`

8. What is a vaccine and how does it work?

   `A vaccine is a weakened or dead version of the pathogen. When the vaccine is injected into a patient, the patient's memory B cells make note of the antigens and can later fight off the real pathogen.`

Critical Thinking Question

1. At the moment, HIV tests are not included in a routine doctor check-up and must be requested by the patient. Should HIV tests become routine and mandated for all sexually active individuals and/or individuals 18 and older? Explain your position.

   `Answers may vary.`

Conclusion

1. Look back at your lab results in Table 6.1 and Table 6.2. What can you conclude about the relationship between the number of sexual partners (without protection) and the risk of receiving a sexually transmitted disease?

   `As the number of partners (without protection) increases, the risk of obtaining a sexually transmitted diseases also increases.`

2. Look back at your lab results in Table 6.3. What do these results indicate about your patients?

Answer Key

Chapter 9
Teacher Guide for Let's Diagnose Them, Lab 4: HIV Test

> Patient #3 is officially diagnosed with HIV. Patients #1, #2, and #4, are HIV negative.

3. Return to your patients' medical charts (Tables 2.1, 2.2, 2.3, and 2.4. pp. 20–23 in Chapter 2, student edition) and complete the section labeled "Lab 4—HIV Test" for each of the four patients. Check off evidence collected from each patient and consider whether or not your original hypothesis is still supported or refuted by evidence.

Reference

Ward's Natural Science Establishment. 1994. Simulated disease transmission lab activity. *www.cteonline.org/portal/default/Resources/Viewer/ResourceViewer?action=2&resid=12395*

Chapter 10
Teacher Guide for Let's Diagnose Them, Lab 5
Lung Capacity

Before the Lesson

- Review the respiratory system with your students. Be sure to discuss:
    - the role of the respiratory system
    - the following organs: *nasal cavity, trachea, epiglottis, bronchi tubes, lungs, alveoli,* and *diaphragm*
    - problems that can arise in the respiratory system: *pneumonia, bronchitis, asthma, hiccups*
    - the term *lung capacity*
- Obtain and set up the following supplies for the Lung Capacity lab.

    For each group of four:
    - 4 inflated balloons of different diameters:
        - Patient #1: Close to 50 cm
        - Patient #2: Close to 38 cm
        - Patient #3: Close to 57 cm
        - Patient #4: Close to 54 cm
    - 1 wind-up measuring tape
    - 1–2 scientific calculators

    Note: Blowing up balloons can be tiresome. Two groups can share a set of balloons and rotate them when done.

Chapter 10
Teacher Guide for Let's Diagnose Them, Lab 5: Lung Capacity

During the Lesson

- Establish roles for each student:
  - Task Manager—reads procedure and ensures that everyone is following proper lab procedures
  - Materials Manager—retrieves and returns materials; cleans materials and table
  - Doctor(s)—completes the lab work (measuring the balloons, performing the calculations). Everyone should assist.
  - Recorder—ensures that the group's data is recorded
- Review the procedure before allowing students to begin. Ask for any clarifying questions.
- Monitor student work. Make sure students understand that they should be comparing the acceptable lung capacity for someone of their patient's height, weight, and gender, with the calculated lung capacity of their patients.

After the Lesson

- Discuss lung capacity results with the class.
- Ensure that students have returned to their patients' medical charts (Tables 2.1, 2.2, 2.3, and 2.4 on pages 20–23 in Chapter 2, student edition) and complete the section labeled "Lab 5—Lung Capacity" for each of the four patients. Make sure they have checked off the evidence collected from each patient and have considered whether or not their original hypothesis is still supported or refuted by the evidence.

Answer Key

Chapter 10
Teacher Guide for Let's Diagnose Them, Lab 5: Lung Capacity

Let's Diagnose Them, Lab 5: Lung Capacity

All four patients have been at Vanguard Hospital for the last two days. Nurses report that all of them are experiencing difficulty breathing. In medical school, you learned the average pair of human lungs can hold about 5 liters or 5,000 cubic centimeters (cc) of air, but only a small amount of this capacity is used during normal breathing (roughly 1,000 cc). Today you will determine the lung capacity of each of your patients. Lung capacity is the maximum amount of air the lungs can hold. Normally, doctors use a spirometer to determine a patient's lung capacity. A spirometer requires the patient to exhale deeply in order to determine if diseases such as asthma, pneumonia, and bronchitis are compromising the patient's respiratory system.

To refresh your memory of lung capacity, let's take a peek at your class notes (Table 7.1) from medical school:

TABLE 7.1. FACTORS THAT MAY INCREASE OR DECREASE LUNG CAPACITY

People With Larger Volumes and Unrestricted Breathing	People With Smaller Volumes and Compromised Breathing
Males	Females
Taller people	Shorter people
Nonsmokers	Smokers
Athletes	Non-athletes
People living at high altitudes	People living at low altitudes
Nonpregnant women	Pregnant women
Healthy weight	Obesity
Normal red blood cells	Sickle cell anemia
Healthy respiratory tracts	Restricted respiratory tracts

As can be seen from the medical school notes, a variety of factors may impact one's lung capacity. Women who are pregnant, for instance, often experience smaller lung volumes since the growing baby pushes up on the diaphragm from the uterus. Similarly, individuals with sickle cell anemia struggle with their breathing since the hemoglobin protein on their RBCs are mutated and only carry half the number of oxygen molecules as a normal RBC. Diabetics also exhibit compromised lung volumes due to high blood sugar levels stiffening the lung tissue and fatty tissue in the abdominal area. And as mentioned previously, those with HIV tend to suffer from opportu-

Chapter 10
Teacher Guide for Let's Diagnose Them, Lab 5: Lung Capacity

Answer Key

nistic infections such as pneumonia, which causes the lungs to fill up with mucous.

Today you will compare each of your patients' lung capacities (from balloon blowing) to the expected lung capacity of someone with the same height, age, and gender. Under normal circumstances, doctors might ask their patients to exhale into a spirometer.

Lab Roles (Fill in Names of Team Members)

1. _____ is the task manager (reads procedure and ensures everyone is following proper protocol).

2. _____ is the materials manager (retrieves and returns materials; cleans materials and table).

3. _____ are the doctors (completes lab work, such as adding chemicals, heating chemicals, and so on; to be completed by more than one group member).

4. _____ is the recorder (ensures the group's data is properly recorded).

Materials

- 4 inflated balloons of different diameters
- 1 wind-up measuring tape
- 1–2 scientific calculators

Procedure

Part I: Patient's Lung Capacity

1. Find the circumference for Patient #1's balloon by wrapping your roll-up ruler around the widest portion of the balloon. Measure the length in centimeters. Record this value in Table 7.2.

2. Repeat step 1 for Patient #2, Patient #3, and Patient #4.

3. Using the formula for circumference, find the radius (r) of the balloon. Plug in the value for C and solve for r. Remember: The value of π is 3.14. Record the value of r in Table 7.2.

Answer Key

Chapter 10
Teacher Guide for Let's Diagnose Them, Lab 5: Lung Capacity

Circumference Equation
$C = 2\pi r$
$r = C/2\pi$

4. Using the radius you just solved for, determine the diameter of the balloon. Plug in the value for r and solve for d. Record the value of d in Table 7.2.

Diameter Equation
$d = 2r$

5. On the x-axis of Figure 7.1, locate the diameter of the balloon in centimeters and follow the number up until it meets the curved line. Then move across, in a straight line, to the vertical y-axis. Approximate the lung volume for your patient. Record the lung volume in Table 7.2.

6. Repeat steps 3–5 for patients #2, #3 and #4.

TABLE 7.2. CALCULATIONS FOR DETERMINING YOUR PATIENTS' LUNG CAPACITIES

	Patient #1	Patient #2	Patient #3	Patient #4
Measure the circumference in centimeters (cm)	50	38	57	54
Calculate the radius in centimeters (cm) $r = C/2\pi$	8	6	9	8.5
Calculate the diameter in centimeters (cm) $d = 2r$	16	12	18	17
Determine your patient's lung capacity (cc) using the graph	2000	1000	3000	2500

DIAGNOSIS FOR CLASSROOM SUCCESS: Making Anatomy • Physiology Come Alive

FIGURE 7.1. DETERMINING LUNG CAPACITY BY A BALLOON'S DIAMETER

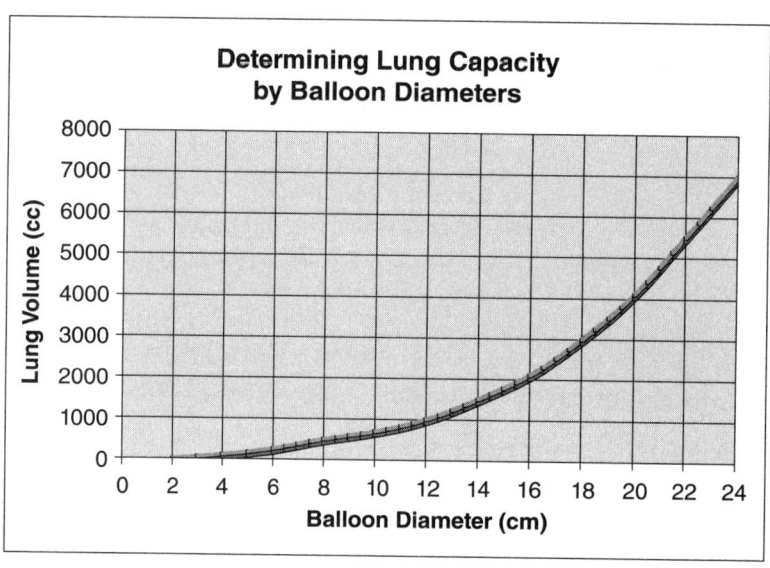

Part II: Acceptable Lung Capacity for Height, Weight, and Gender

Research has shown the capacity of a person's lungs *should* be proportional to the surface area of his or her body. To find the surface area of your patients, you will need to know the height, weight, and gender of each, which are listed in their medical records. There are a couple of different ways to calculate mathematically a person's body surface area and estimate their *acceptable* lung capacity, mathematically.

To determine the acceptable lung capacity of your patients, enter their heights, weights, and gender values into the Body Surface Area equation below. Record this value in Table 7.3. *Note:* Height must be in centimeters and weight must be in kilograms. Your head doctor (the teacher) has completed these conversions for you already. Use the **bold** values for your calculations.

$$\text{Body Surface Area} = \sqrt{(\text{height in cm} \times \text{weight in kg})/3600}$$

$$\text{If } Female = \text{Body Surface Area Value} \times 2000$$

$$\text{If } Male = \text{Body Surface Area Value} \times 2500$$

Answer Key

Chapter 10
Teacher Guide for Let's Diagnose Them, Lab 5: Lung Capacity

TABLE 7.3. DETERMINING PATIENTS' ACCEPTABLE LUNG CAPACITIES

	Sex	Height (ft → cm)	Weight (lbs → kg)	Acceptable Lung Capacity (cc) (Show math!)
Patient #1	F	5'6" → **167.7 cm**	200 lbs → **90.7 kg**	4,109.8
Patient #2	F	5'2" → **157.5 cm**	120 lbs → **54.5 kg**	3,085.4
Patient #3	M	6'2" → **188.0 cm**	140 lbs → **63.5 kg**	4,552.5
Patient #4	M	5'8" → **172.7 cm**	155 lbs → **70.3 kg**	4,591.1

7. Compare the results you obtained in Tables 7.2 and 7.3 and complete Table 7.4.

TABLE 7.4. COMPARISON OF EACH PATIENT'S ACTUAL LUNG CAPACITY TO HIS OR HER ACCEPTABLE LUNG CAPACITY

	Patient #1	Patient #2	Patient #3	Patient #4
How does your patient's lung capacity compare to the acceptable lung capacity of someone with the same height, weight, and gender (higher, lower, or similar)?	Lower	Lower	Lower	Lower

Recall Questions

1. What is the role of the respiratory system?

   The respiratory system allows oxygen to enter the body and lets carbon dioxide out.

2. When a person exhales, what happens to the diaphragm? What happens during an inhale?

   When a person exhales, the diaphragm relaxes and pushes upward, decreasing the space of the lungs and pushing air

Chapter 10
Teacher Guide for Let's Diagnose Them, Lab 5: Lung Capacity

Answer Key

outward. During an inhale, the diaphragm contracts and flattens, allowing air to come inward.

3. What happens in the alveoli?

   The alveoli are tiny air sacs surrounded by capillaries. Diffusion of gases occurs here. Inhaled oxygen moves from the lungs into the circulatory system, and carbon dioxide exits the circulatory system and enters the lungs to be exhaled.

4. Why would a pregnant person have a lower lung capacity?

   A pregnant woman would have a lower lung capacity because the growing baby pushes upward on the diaphragm, decreasing the volume of the lungs.

5. Why might a person with sickle cell anemia have difficulty breathing?

   A person with sickle cell anemia would have difficulty breathing because their red blood cells carry half the normal amount of red blood cells due to the mutated hemoglobin protein.

6. Why would someone who has diabetes have difficulty breathing?

   A person who is diabetic often has a difficult time breathing because excess sugar can cause stiffening of the lung tissue, making it less elastic.

7. Why would someone who has HIV have difficulty breathing?

   Those with HIV are more susceptible to opportunistic infections like pneumonia, which can impact one's ability to breathe.

Critical Thinking Question

1. Asthma is a respiratory disease characterized by restricted airflow, resulting in difficulty breathing and a variety of other chest-related symptoms. Asthma attacks can be brought on by allergies, but are often brought on by poor air quality. In 2012 the U.S. Centers for Disease Control and Prevention stated that asthma cases rose from 7.3% in 2001 to 8.4% in 2010 and were higher among children, than adults, and among multiple-race, black, and American Indian or Alaska Native persons than white persons. In particular, low-income urban areas tended to have higher asthmatic cases. Considering your knowledge of respiratory health, is it the government's responsibility to make asthma prevention a priority or

should business and factories be held accountable for improving outdoor air quality? Create a proposal that would satisfy the needs of urban residents, government officials, and big businesses.

`Answers may vary.`

Conclusion

1. Look back at your lab results in Table 6.3. What could these results indicate about your patients?

   `All four patients' breathing is being compromised by their health condition (diabetes, pregnancy, sickle cell anemia, and HIV).`

2. Return to your patients' medical charts (Tables 2.1, 2.2, 2.3, and 2.4. pp. 20–23 in Chapter 2, student edition) and complete the section labeled "Lab 5—Lung Capacity" for each of the four patients. Check off evidence collected from each patient and consider whether or not your original hypothesis is still supported or refuted by evidence.

References

Akinbami, L. J., J. E. Moorman, C. Bailey, H. S. Zahran, M. King, C. A. Johnson, and X. Liu. 2012. Trends in asthma prevalence, health care use, and mortality in the United States, 2001–2010. Centers for Disease Control and Prevention, NCHS Data Brief: 94 (May). *www.cdc.gov/nchs/data/databriefs/db94.htm*

The Biology Corner. Measuring lung capacity. *www.biologycorner.com/worksheets/lungcapacity.html*

Diabetes in Control. 2008. Reduced lung capacity accelerates with diabetes. Diabetes Care 410 (April 8). *http://173-203-204-222.static.cloud-ips.com/articles/diabetes-news/5637.*

Gingo, M. R., and M. P. George. 2010. Pulmonary function abnormalities in HIV infected patients during the current antiretroviral therapy era. *American Journal of Respiratory and Critical Care Medicine* 182 (6): 6.

Strategic Education Research Partnership. 2010. Asthma: More than a medical problem? *http://wg.serpmedia.org/pdf_1BTE/WG1BTE_individual_units/full_gray/WG1BTE_gray_Part14.pdf*

WebMD. Spirometry and other lung function tests. *www.webmd.com/lung/lung-function-tests.*

Chapter 11
Teacher Guide for Let's Diagnose Them, Lab 6
Hormone Test

Before the Lesson

- Review the male and female reproductive systems with your students. Be sure to discuss:
    - the role of reproduction
    - the following male sex cell and organs: *sperm, testes, scrotum, vas deferens, prostate, urethra, penis*
    - the following female sex cell and organs: *egg, ovaries, fallopian tubes, uterus, cervix, vagina*
    - problems that may arise in the reproductive track: *ovarian cysts, cervical cancer, testicular cancer, prostate cancer*
    - changes that occur in mother and child during first, second, and third trimesters
    - the medical terms: *pregnancy, miscarriage, ectopic pregnancy, preeclampsia*
    - various forms of *birth control*
- Review the endocrine system with your students. Be sure to discuss:
    - the role of the endocrine system and hormones
    - the following glands: *pituitary, hypothalamus, thyroid, adrenal, testes, ovaries, pancreas, liver*
    - the following hormones: *testosterone, estrogen, progesterone, human chorionic gonadotropin (hCG), insulin, glycogen*
    - the glycemic index (again)
    - problems that may arise: *type 1 and 2 diabetes*

Chapter 11
Teacher Guide for Let's Diagnose Them, Lab 6: Hormone Test

- Obtain and set up the following supplies for the hCG Pregnancy test:

    For the entire class:
    - 1 250 ml sample of Patient #1 urine: water and yellow food coloring
    - 1 250 ml sample of Patient #2 urine: water, yellow food coloring and vinegar

    For each group of four:
    - 2 test tubes
    - 1 test tube rack
    - "Pregnancy" test (blue litmus paper)
    - Students should obtain 3 ml urine samples from Patients #1 and #2

During the Lesson

- Establish roles for each student:
    - Task Manager—reads procedure and ensures that everyone is following proper lab procedures
    - Materials Manager—retrieves and returns materials; cleans materials and table
    - Doctor(s)—completes the lab work (i.e., adding chemicals)
    - Recorder—ensures that the group's data is recorded
- Review the procedure before allowing students to begin. Ask for any clarifying questions.
- Monitor student work. Ensure that students understand how to read the glucose tolerance test results.

After the Lesson

- Discuss pregnancy and glucose tolerance test results with the class.
- Ensure that students have returned to their patients' medical charts (Tables 2.1, 2.2, 2.3, and 2.4 on pages 20–23 in Chapter 2, student edition) and complete the section labeled "Lab 6 – Hormone Tests" for each of the four patients. Make sure they have checked off the evidence collected from each patient and have considered whether or not their original hypothesis is still supported or refuted by the evidence.

Answer Key

Chapter 11
Teacher Guide for Let's Diagnose Them, Lab 6: Hormone Test

Let's Diagnose Them, Lab 6: Hormone Test

Your patients have been in this hospital for three days. By now, you have most likely verified a diagnosis for your male patients, Patients #3 and #4. You also might suspect that one of your female patients is pregnant and that the other is diabetic.

Every doctor knows the endocrine system produces hormones that help regulate the body's internal organs. Hormones are chemical messengers sent throughout the bloodstream. When too little or too much of a hormone is produced, it is often an indication that something is wrong.

In medical school (Table 8.1), you learned the following about two hormones produced in the body:

TABLE 8.1. MEDICAL SCHOOL NOTES REGARDING TWO HORMONES: HCG AND INSULIN

	Increased levels could indicate …	Decreased levels could indicate …
Human Chorionic Gonadotrophin (hCG)	• Pregnancy only	• Not pregnant
Insulin	• Drugs such as corticosteroids, levodopa, and oral contraceptives • Fructose or galactose intolerance • Excessive exercising	• Diabetes • Pancreatic diseases such as chronic pancreatitis and pancreatic cancer

To determine if someone is diabetic, a doctor must provide patients with a glucose tolerance test. This test records how quickly sugar is cleared from the blood stream. The test is most frequently used to determine if a person is diabetic. The patient in question is required to fast 8 to 14 hours before they take the test. Only water is allowed. The patient is then given a glucose solution to drink. Blood is drawn at different intervals, and glucose levels are measured each hour. The glucose levels following the 2-hour mark are the most critical in determining if a person is diabetic. Glucose levels above 200 mg/dl show that insulin levels are low, suggesting diabetes.

Now that you have reviewed your medical school knowledge, let's diagnose these patients!

Chapter 11
Teacher Guide for Let's Diagnose Them, Lab 6: Hormone Test

Answer Key

Lab Roles (Fill in Names of Team Members)

1. _____ is the task manager (reads procedure and ensures everyone is following proper protocol).

2. _____ is the materials manager (retrieves and returns materials; cleans materials and table).

3. _____ are the doctors (completes lab work, such as adding chemicals, heating chemicals, and so on; to be completed by more than one group member).

4. _____ is the recorder (ensures the group's data is properly recorded).

Materials

- 2 test tubes
- 1 test tube rack
- blue litmus paper ("pregnancy" test)
- 3 ml urine samples from Patients #1 and #2 (retrieve from your teacher)

Procedure

Part I: Pregnancy Test

1. Label two test tubes *Patient #1* and *Patient #2*.

2. Obtain a 3 ml urine sample from both patients. Be sure to place them into their designated test tubes.

3. Dip the BLUE litmus paper ("pregnancy test") into the urine sample of Patient #1. If the pregnancy test does not change colors, as indicated in Table 8.2, the patient has normal levels of hCG in their urine and is not pregnant. If the pregnancy test changes to a PINK color, the patient has high levels of hCG in their urine and is therefore pregnant.

4. Record data for Patient #1 in Table 8.3.

5. Repeat steps 3–4 for Patient #2.

Answer Key

Chapter 11
Teacher Guide for Let's Diagnose Them, Lab 6: Hormone Test

TABLE 8.2. MEANING OF HCG INDICATOR RESULTS

If color does not change, it indicates ...	If color changes, it indicates ...
No pregnancy	Pregnancy

TABLE 8.3. RESULTS OF PATIENT #1'S AND #2'S PREGNANCY TESTS

	Patient #1	Patient #2	Patient #3	Patient #4
Pregnancy test • hCG Present?			male (no pregnancy test done)	male (no pregnancy test done)

Part II: Glucose Tolerance Test

1. Interpret the Glucose Tolerance Test results (Figure 8.1) for all four patients.
2. Use your medical school notes (Table 8.4) to determine which patients are normal or diabetic.
3. Record results in Table 8.5.

TABLE 8.4. MEDICAL SCHOOL NOTES REGARDING NORMAL AND DIABETIC TEST RESULTS FOR A GLUCOSE TOLERANCE TEST

Glucose levels	Normal		Diabetic	
Venous Blood Plasma	Fasting	2 hrs	Fasting	2 hrs
(mg/dl)	<110	<140	>126	>200

DIAGNOSIS FOR CLASSROOM SUCCESS: Making Anatomy & Physiology Come Alive

Chapter 11
Teacher Guide for Let's Diagnose Them, Lab 6: Hormone Test

FIGURE 8.1. RESULTS FOR EACH PATIENT'S GLUCOSE TOLERANCE TEST

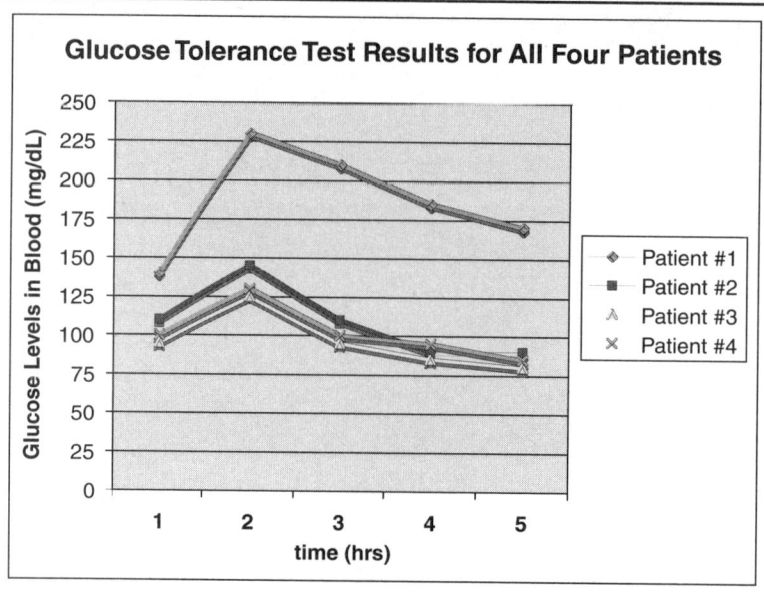

TABLE 8.5. RESULTS FOR EACH PATIENT'S GLUCOSE TOLERANCE TEST

	Patient #1	Patient #2	Patient #3	Patient #4
Glucose level after two hours: Above 200 mg/dl? Between 145–200 mg/dl? Below 145 mg/dl?	Above	Between	Below	Below
Insulin level must be: Low? Borderline? Normal?	Low	Borderline	Normal	Normal

Recall Questions

1. What is the role of the reproductive system?

   The reproductive system is responsible for producing offspring.

2. What is the role of the endocrine system?

   The endocrine system is responsible for producing hormones that help regulate the body.

3. Why is hCG only found in pregnant women?

   hCG is produced by the placenta, after conception.

Answer Key

Chapter 11
Teacher Guide for Let's Diagnose Them, Lab 6: Hormone Test

4. What is the role of insulin?

   `Insulin helps lower glucose levels in the blood by helping it enter cells of the body.`

5. What organ produces insulin?

   `The pancreas produces insulin.`

6. Explain how insulin and glycogen are examples of a feedback loop. In other words, how do they help maintain homeostasis in one's body?

   `When sugar levels are high, the pancreas releases insulin to help lower sugar levels. Conversely, when sugar levels in the blood are too low, the liver releases glycogen, which helps release stored sugar from muscle cells back into the blood stream.`

7. What is the difference between type 1 and type 2 diabetes?

   `Type 1 Diabetes often occurs in children. It is often known as juvenile diabetes. Those with type 1 diabetes do not produce insulin. Type 2 diabetes usually occurs later in life due to poor nutrition and exercise. Obesity is also a risk factor for type 2 diabetes. The insulin of those with type 2 diabetes is produced minimally or is inefficient at lowering glucose levels in the blood stream.`

Critical Thinking Question

1. According to Guttmacher Institute's "State Policies in Brief" (2012), "37 states require parental involvement in a minor's decision to have an abortion" (p. 1). The term *parental involvement* however, ranges in definition. For some states, it refers to parental consent or others' notification, and for some, a notarized document. Additionally, the need for parental involvement varies case by case (i.e., medical emergency, abuse/assault/incest/neglect, and so on). Suppose you were the doctor of a 16-year-old pregnant patient in a state that did *not* mandate parental involvement, what medical advice might you provide to your patient to help her make an informed decision moving forward?

   `Answers may vary.`

Conclusion

1. Look back at your lab results in Table 8.3 and Table 8.5. What could these results indicate about your patients?

Chapter 11
Teacher Guide for Let's Diagnose Them, Lab 6: Hormone Test

Answer Key

```
Since hCG was found in Patient #2's urine, she can
officially be diagnosed as pregnant. Additionally, her
glucose tolerance test results indicate that her sugar
levels are on the border of normal and diabetic, so she
could be suffering from gestational diabetes. Patient #1
is not pregnant, but her glucose tolerance test results
indicate that her sugar level is high. Therefore, she has
low insulin and is probably developing type 2 diabetes,
given her medical record and other symptoms. Patient #3
and #4 were not tested for pregnancy since they are males.
Additionally, their glucose tolerance test results appear
normal.
```

2. Return to your patients' medical charts (Tables 2.1, 2.2, 2.3, and 2.4. pp. 20–23 in Chapter 2, student edition) and complete the section labeled "Lab 6—Hormone test" for each of the four patients. Check off evidence collected from each patient and consider whether or not your original hypothesis is still supported or refuted by evidence.

References

American Association for Clinical Chemistry. Lab tests online—Insulin: The test. *http:// labtestsonline.org/understanding/analytes/insulin/tab/test*

American Pregnancy Association. Human chorionic gonadotropin (HCG): The pregnancy hormone. *www.americanpregnancy.org/duringpregnancy/ hcglevels.html*

Guttmacher Institute. 2013. Parental involvement in minors' abortions. *State policies in brief. www.guttmacher.org/statecenter/spibs/spib_PIMA.pdf*

National Institutes of Health. MedLine plus—Glucose tolerance test. *www.nlm.nih. gov/ medlineplus/ency/article/003466.htm*

World Health Organization. 2006. Definition and diagnosis of diabetes mellitus and Intermediate hyperglycemia. *www.who.int/diabetes/publications/ Definition%20 and%20diagnosis%20of%20diabetes_new.pdf*

Chapter 12
Teacher Guide for Emergency! Lab 7
Performing Surgery

Before the Lesson

- Review the body systems covered within this book:
    - Excretory
    - Digestive
    - Circulatory
    - Immune
    - Respiratory
    - Reproductive
    - Endocrine
- Introduce the concept of organ donations. Consider showing students various statistics.
- Review the lab safety and dissection protocol with your students. Be sure to discuss:
    - The dissection tools and their functions
    - The proper disposal of waste materials
    - The clean-up procedure
- Obtain and set up the following supplies for the dissection:

    For each group of four:
    - 1 rat
    - 1 dissection tray
    - 1 scalpel/dissection scissor

Chapter 12
Teacher Guide for Emergency! Lab 7: Performing Surgery

- several dissection pins
- 1 dropper
- gloves for each person handling the rat
- goggles for each team member
- 1 biology textbook for reference

During the Lesson

- Establish roles for each student:
    - Task Manager—reads procedure and ensures that everyone is following proper lab procedures
    - Materials Manager—retrieves and returns materials; cleans materials and table
    - Doctor(s)—completes the lab work (i.e., make incisions); may be done by more than one person.
    - Recorder—ensures that the group's data is recorded
- Review the procedure before allowing students to begin. Ask for any clarifying questions.
- Monitor student work. *Note:* Some students might get queasy during this lab. Consider allowing students to perform a virtual lab dissection online.

After the Lesson

- Discuss observations with the class.
- Now might be a good time to implement a third assessment on the respiratory system, reproductive system, and endocrine system. Some teachers prefer to assess all the systems taught throughout this narrative.

Chapter 12
Teacher Guide for Emergency! Lab 7: Performing Surgery

Emergency! Lab 7: Performing Surgery

Fortunately, you have been taking great care of your patients. However, a fifth patient (represented by the rat) has just been rushed into the emergency room at Vanguard Hospital. Doctors have been working on this patient, tirelessly, but were unsuccessful reviving him/her. The patient has indicated that he/she is an organ donor. Doctors in the operating room have called upon you to perform the donation surgery. Given your knowledge of anatomy and physiology from medical school and your work at Vanguard Hospital, it will be your job to identify and remove various organs from the patient's body.

Lab Roles (Fill in Names of Team Members)

1. _____ is the task manager (reads procedure and ensures everyone is following proper protocol).

2. _____ is the materials manager (retrieves and returns materials; cleans materials and table).

3. _____ are the doctors (completes lab work, such as adding chemicals, heating chemicals, and so on; to be completed by more than one group member).

4. _____ is the recorder (ensures the group's data is properly recorded).

Materials (Per Group of Four)

- 1 rat
- 1 dissection tray
- 1 scalpel/dissection scissor
- several dissection pins
- 1 dropper
- gloves for each person handling the rat
- goggles for each team member
- 1 biology textbook for reference

Chapter 12
Teacher Guide for Emergency! Lab 7: Performing Surgery

Procedure

1. Make sure all doctors (students) are wearing proper safety gear (gloves, goggles)
2. Lay the rat on its back.
3. Pierce the rat's abdomen with the scalpel/dissection scissor. Cut vertically (Figure 9.1) from the top of the abdomen area to the pelvic region.
4. At the top of the abdomen and at the bottom of the abdomen, cut horizontally. There should now be two flaps that open like a book. Pin these flaps back with the dissection pins.

FIGURE 9.1. PREPARING FOR RAT DISSECTION

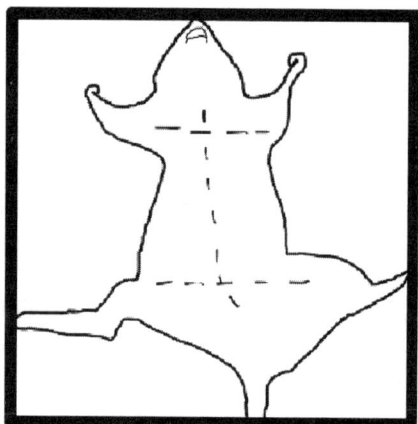

5. Right now you are looking inside the rat's abdomen area.

- **Question 1:** What organ takes up most of the space in the rat's abdomen?

    `Liver`

- **Question 2:** Identify the stomach. Sketch it below.

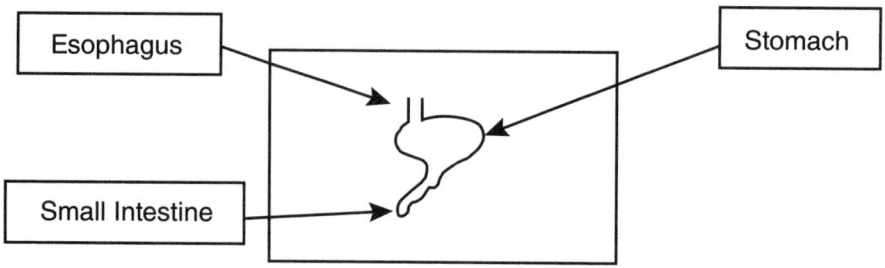

Answer Key

Chapter 12
Teacher Guide for Emergency! Lab 7: Performing Surgery

- **Question 3:** What tube is leading toward the stomach?

    `Esophagus`

- **Question 4:** What tube is connected to the bottom of the stomach?

    `Small intestine`

6. Cut out the stomach. Open it up.

- **Question 5:** What does the stomach look like inside? Describe it.

    `Several ridges / Wrinkly`

7. Gently pull out the small intestine.

- **Question 6:** How long is the small intestine? Measure its length in centimeters.

    `Answers will vary but it should be much longer than the large intestine.`

- **Question 7:** What is the job of the small intestine?

    `The small intestine absorbs nutrients into the bloodstream.`

8. Gently pull out the large intestine.

- **Question 8:** How long is the large intestine? Measure its length in centimeters.

    `Answers will vary but it should be much shorter than the small intestine.`

- **Question 9:** What is the job of the large intestine?

    `The large intestine reabsorbs water.`

9. Open the large intestine.

- **Question 10:** What is inside of it?

    `Undigested material (feces)`

10. At this point, you have already removed a large portion of the organs in the abdominal cavity of the rat. Located toward the back of the rat are two beanlike structures, called the kidneys. See if you can find them.

- **Question 11:** What is the function of the kidneys?

    `The kidneys filter blood.`

Chapter 12
Teacher Guide for Emergency! Lab 7: Performing Surgery

Answer Key

11. Open up one of the kidneys.

- **Question 12:** What does the kidney look like inside?

    `There are lots of blood vessels inside the kidney.`

12. The kidneys are attached to two tubes called the ureter. The ureter connects to the bladder.

- **Question 13:** What is the role of the bladder?

    `The bladder stores urine.`

13. Since you are now in the pelvic region of the rat, see if you can determine the gender of your rat.

- **Question 14:** What gender is your rat? (`Answers will vary`) How do you know? What did you find—or not find—that may support your answer?

    `Male rats will have testicles and a penis.`

14. Now change gears. Let's concentrate on the upper region of the rat that hasn't been exposed. At the very top there should be a membrane-like muscle separating the upper abdomen from the chest region of the rat.

- **Question 15:** What is the name of this muscle that is located below the ribs that spans the width of the rat?

    `The diaphragm`

15. Make a vertical cut through the chest of the rat. You will probably need to use a little bit more force to break through the ribs. Once cut, open up the chest cavity. Here you should see the lungs and the heart.

- **Question 16**: Sketch the structure of the lungs and heart below.

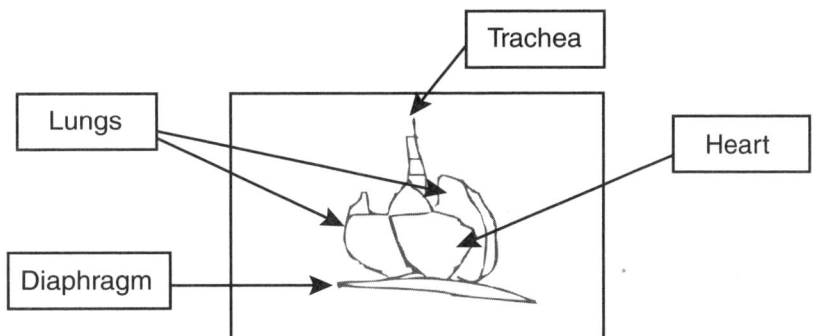

- **Question 17**: Explain how the heart and lungs work together to help your body function properly.

  As oxygen (O_2) comes into the lungs, the pulmonary vein brings the O2 to the left atrium of the heart, to the left ventricle of the heart, and releases it to the rest of the body by way of the aorta. Once the O_2 reaches the cells, the cells release the waste product, carbon dioxide. Carbon dioxide (CO_2) enters the right atrium of the heart via the superior and inferior vena cava and heads to the right ventricle. The pulmonary artery will bring that CO_2 back to the lungs where it can be released from the body. This cycle continues.

16. If you glance around the rat, you probably notice hot pink blood vessels and blue blood vessels. The rats have been injected with a serum to help identify the arteries and veins.

    - **Question 18**: What is the job of the arteries?

      Arteries carry oxygenated blood away from the heart and toward cells.

    - **Question 19**: What is the job of the veins?

      Veins carry deoxygenated blood toward the heart and away from cells.

    - **Question 20**: Use your knowledge of the circulatory system to explain why the hot pink and blue blood vessels are misleading.

      Arteries and veins are often depicted as red and blue in textbooks to help differentiate the pathway of blood. However, veins are not blue; they are a deep red color. Studies indicate that there are four factors influencing the color of blood in these two vessels: (1) wavelengths

Chapter 12
Teacher Guide for Emergency! Lab 7: Performing Surgery

Answer Key

deflected and absorbed by a person's skin tissue (2) the oxygen levels in blood (3) the diameter and depth of the blood vessels, and (4) visual perception.

17. Try to identify the pipe leading from the mouth to the lungs.

- **Question 21:** What is the name of this pipe?

    The trachea

18. Job well done! Make sure you clean up and wipe down your area. Nurses were able to preserve Patient #5's organs in an ice-cold preservative solution and have already packed them into sterile containers. These containers contain an icy slush mixture that will help prevent cell deterioration.

Recall Questions

1. Why are dissections useful?

    Dissections are useful because we can learn the structure and function of organs.

2. Why do you think rats were selected as Patient #5, opposed to frogs?

    Like humans, rats are mammals. Their body is constructed similarly, but on a smaller scale. Frogs are amphibians; therefore their body systems (i.e., respiratory, reproductive) do not look or work the same way.

3. What did you find challenging during this lab?

    Answers will vary.

4. What did you like most about this lab?

    Answers will vary.

Critical Thinking Question

1. Although there have been several advancements in technology and medicine, the demand for organs far surpasses the number of organ donors. In order to identify oneself as an organ donor in the United States, a person needs to "opt-in," meaning he or she is not considered an organ donor until he or she takes concrete action to be one. In several European countries, however, a person is considered an organ donor until he or she "opts out" and illustrates an unwillingness to donate. Given what you have learned about the body systems and how easily homeostasis can be disrupted, suppose the United States considered an "opt out" program, would you vote for or against it? Justify your position.

    Answers will vary.

Conclusion
Time to get back to your four original patients!

References
Kienle, A., and L. Lilgle. 1996. Why do veins appear blue? A new look at an old question. *Applied Optics* 35 (7): 1151–1160.

TransWeb.org. How are donated organs preserved and transported? University of Michigan Transplant Center. *www.transweb.org/faq/q23.shtml*

Chapter 13
Teacher Guide for The Ominous Phone Call and Evaluating the Docs

Before the Lesson

- Make sure your students have access to computers so that they can research an appropriate treatment and prognosis for their patients' health conditions.

- Determine what kind of final assessment you would like to measure your students with:
    - Visual/Oral PowerPoint
        - PowerPoint (for all four patients' cases)
        - Public Service Announcement (for one patient's case)
    - Written
        - Essay #1
        - Essay #2
    - Exam

Note: Both the teacher and student editions contain instructions for visual/oral and written assessments only. There is no sample exam or review sheet in either edition.

- Encourage students to collaborate during their visual/oral assessments. Programs that help promote collaboration among students and educator include:
    - GoogleDocs
    - Wikispaces
    - Prezi

Chapter 13
Teacher Guide for The Ominous Phone Call and Evaluating the Docs

- It is highly recommended that you set aside a day for students to present their work to other science teachers and/or outside evaluators. Recruit doctors and medical students to help assess your students' understanding. Students really "step it up" when they have an authentic audience to address. If this is too difficult to organize, then in-class presentations will suffice. *Note:* Since all students completed the same tasks, with the same characters and health conditions, the presentations may become somewhat tedious for one class audience.

During the Lesson

- Monitor student progress. Ensure that all students understand how they will be assessed.

- When assessing students during the presentation and written portion of this project, write each student's name on the lines under the heading "Student." Use the rubric to determine if each student is Novice, Competent, or an Expert for each of the standards listed. Place an "x" on the lines beside their names to indicate their level.

After the Lesson

- Be sure to include feedback to your students so that they can improve their presentation and scientific skills in the future.

- Celebrate your students' successes!

The Ominous Phone Call

Filling out a Prescription

After several days of testing, you have finally developed a diagnosis for each of your four patients. Before releasing them from the hospital you *must* complete the medical charts (Tables 2.1, 2.2, 2.3, and 2.4 on pages 20–23 in Chapter 2, student edition) and provide the pharmacist with the following pieces of information under the sections "Diagnosis," "Treatments," and "Prognosis":

1. Results from each of the six tests and a final **diagnosis** of your patient's condition.

2. **Treatments** available for your patient's condition, and

3. A **prognosis**. A prognosis is a medical report dictating a physician's view on a case. It often denotes the chance of a patient's recovery and the doctor's prediction of how that patient will progress. For instance, one might note if the condition is short-term, long-term, fatal, and so on.

Chapter 13
Teacher Guide for The Ominous Phone Call and Evaluating the Docs

Evaluating the Docs

The head doctor (your teacher) would like your team to present your findings for evaluation and to develop a presentation that encompasses the following information. Be sure to include appropriate data tables and graphs to explain their conditions. When developing your PowerPoint or Public Service Announcement, be sure to collaborate with your fellow doctors. Programs you might want to consider are:

- Google Docs
- Wikispaces
- Prezi

Part I: Oral/Visual Assessment Options

Option 1: PowerPoint (for all four patient cases)

Slide 1: Introduction

a) Introduce yourselves

b) The goal of the project

c) An overview of the four diseases

Slide 2: Lab 1—Urinalysis

a) What was the purpose of this lab?

b) What system of the body is it associated with?

c) How was it done?

d) What could results indicate?

Slide 3: Lab 2—Digestive By-Products and BMI Analysis

a) What was the purpose of this lab?

b) What system of the body is it associated with?

c) How was it done?

d) What could results indicate?

Answer Key

Chapter 13
Teacher Guide for The Ominous Phone Call and Evaluating the Docs

Slide 4: Lab 3—Blood Smears

  a) What was the purpose of this lab?

  b) What system of the body is it associated with?

  c) How was it done?

  d) What could results indicate?

Slide 5: Lab 4—HIV Test

  a) What was the purpose of this lab?

  b) What system of the body is it associated with?

  c) How was it done?

  d) What could results indicate?

Slide 6: Lab 5—Lung Capacity

  a) What was the purpose of this lab?

  b) What system of the body is it associated with?

  c) How was it done?

  d) What could results indicate?

Slide 7: Lab 6—Hormone Test

  a) What was the purpose of this lab?

  b) What system of the body is it associated with?

  c) How was it done?

  d) What could results indicate?

Slide 8: Patient #1 Diagnosis

  a) Describe her medical records symptoms.

  b) What was your original hypothesis and why?

  c) Diagnosis: What health condition do you think she has now and why?
    Use prescription pad for help.

Chapter 13
Teacher Guide for The Ominous Phone Call and Evaluating the Docs

Answer Key

Slide 9: Patient #1 Health Condition

a) What could have caused her condition?

b) What are the symptoms of this health condition?

c) What are potential treatments?

d) What is the prognosis?
  Use prescription pad and medical school notes for help.

Slide 10: Patient #2 Diagnosis

a) Describe her medical records symptoms.

b) What was your original hypothesis and why?

c) Diagnosis: What health condition do you think she has now and why?
  Use prescription pad for help.

Slide 11: Patient #2 Health Condition

a) What could have caused her condition?

b) What are the symptoms of this health condition?

c) What are potential treatments?

d) What is the prognosis?
  Use prescription pad and medical school notes for help.

Slide 12: Patient #3 Diagnosis

a) Describe his medical records symptoms.

b) What was your original hypothesis and why?

c) Diagnosis: What health condition do you think he has now and why?
  Use prescription pad for help.

Slide 13: Patient #3 Health Condition

a) What could have caused his condition?

b) What are the symptoms of this health condition?

c) What are potential treatments?

d) What is the prognosis?
  Use prescription pad and medical school notes for help.

NATIONAL SCIENCE TEACHERS ASSOCIATION

Answer Key

Chapter 13
Teacher Guide for The Ominous Phone Call and Evaluating the Docs

Slide 14: Patient #4 Diagnosis

   a) Describe his medical records symptoms

   b) What was your original hypothesis and why?

   c) Diagnosis: What health condition do you think she has now and why?
*Use patient medical charts for help

Slide 15: Patient #4 Health Condition

   a) What could have caused her condition?

   b) What are the symptoms of this health condition?

   c) What are potential treatments?

   d) What is the prognosis?
*Use patient medical charts and medical school notes for help.

Slide 16: Conclusion
Each doctor should have a closing statement that entails one or more of the following:

   a) What have you learned from doing this project?

   b) How does the subject matter relate to your life or community at large?

   c) What advice might you give others regarding their health?

   d) What did you find most interesting?

   e) What questions did this project raise for you?

Option 2: Public Service Announcement (PSA)—(for one patient case)

Create a public service announcement for one of your patients' health conditions that addresses the questions listed below. Be sure to include evidence to support your claims.

1. What condition is the PSA for?

2. How does one get this condition?

3. What are some of the earlier symptoms?

Chapter 13
Teacher Guide for The Ominous Phone Call and Evaluating the Docs

4. Who should get tested for this health condition, and at what point?

5. How is one diagnosed with this health condition? For instance, what lab tests will they have to do to verify their condition?

6. What are the later symptoms and why? How is the body affected by this condition?

7. Can this condition be treated, cured, or managed? If so, what treatments, cures, or management tips are available?

8. What is the prognosis for a patient who is treated (versus untreated) for your condition?

Part II. Written Assessment Options

To renew your license, you must submit a written portion summarizing your work. Select *one of the two* questions below and write a 1–2 page paper using evidence from your patients' symptoms, the labs performed, and your medical school knowledge to support your ideas.

Option 1

The human body is made up of several different systems. Each system has a separate function, but as you've learned, many of them work together. Below is a list of systems we have discussed in class. Select *two* systems from the list below. Explain how they interact with one another to help keep a living organism healthy and alive. Be sure to cite examples from this project.

- Excretory System
- Digestive System
- Circulatory System
- Immune System
- Respiratory System
- Reproductive System
- Endocrine System
- Nervous System (*Note: The nervous system was not covered in this project*)
- Musculoskeletal System (*Note: The musculoskeletal system was not covered in this project.*)

Answer Key

Chapter 13
Teacher Guide for The Ominous Phone Call and Evaluating the Docs

Option 2

One's genes and/or lifestyle choices can disrupt homeostasis. Give concrete examples for each of the *two* factors listed above (genes and lifestyle choices), and explain how they may interfere with the health of a human being. Be sure to cite specific examples from this project.

a) Who was affected?

b) What condition did he/she suffer from?

c) How did this condition come about in him/her?

d) What symptoms did he or she experience?

e) What lab tests helped diagnose this condition? Explain how the results were determined.

f) Are any treatments or cures available? If so, what can the patient do to monitor or control his or her condition?

Chapter 13
Teacher Guide for The Ominous Phone Call and Evaluating the Docs

Answer Key

Mystery Diagnosis Rubric for Evaluating the Docs

Doctors are constantly being evaluated on their performance. Great ones may renew their license and are often recommended by patients through referrals, while poor-performing doctors, may lose their license to practice. Today, the head doctor (your teacher) is going to evaluate your performance as a doctor. He or she will determine if you are performing competently or whether or not you may need to be re-evaluated. Below is the rubric for your evaluation.

Using Evidence: The "doctor" can explain the purpose and the methods of the tests he or she performed to make a diagnosis. He/she incorporates all the data to draw valid conclusions.

Student

_____ : <--->

_____ : <--->

_____ : <--->

_____ : <--->

Novice	Competent	Expert
The "doctor" cannot explain the lab in a way that the evaluator can understand the purpose, procedures, nor the results.	The "doctor" appears more confident about the purpose, procedure, and results of some labs over others.	The evaluator has a *clear understanding* of the lab's purpose, procedure, and results based on the "doctor's" explanation.

NATIONAL SCIENCE TEACHERS ASSOCIATION

Answer Key

Chapter 13
Teacher Guide for The Ominous Phone Call and Evaluating the Docs

Making Connections: The "doctor" is able to make connections between the knowledge of the body systems, research on the health conditions, and evidence gathered from each lab.

Student

_____ : <-->

_____ : <-->

_____ : <-->

_____ : <-->

Novice	Competent	Expert
The "doctor" does not have a thorough understanding of the body systems and/or diseases; information is inaccurate and/or has difficulty answering questions presented by evaluator.	The "doctor" has a decent understanding of each body system and condition, and sometimes struggled with the questions presented by the evaluator. Most of the information is accurate.	The "doctor" has a *thorough understanding* of each body system and condition; information is accurate, and therefore it is *easy* to address questions posed by the evaluator.

Seeking Significance: The "doctor" understands the importance of these topics and how it affects his or herself, people within his or her community, and those worldwide. He or she can provide recommendations on how to prevent, treat, or manage the conditions diagnosed.

Student

_____ : <-->

_____ : <-->

_____ : <-->

_____ : <-->

Novice	Competent	Expert
The "doctor" does not clearly indicate ways in which this topic is relevant. Lacks suggestions for others on how to prevent, treat, or manage the condition.	The "doctor" is limited in his/her understanding of how this topic is relevant and is limited in his/her suggestions on how to prevent, treat, or manage the condition.	The "doctor" clearly indicates ways in which this topic is relevant and makes suggestions for others to prevent, treat, or manage the disease or condition.

Chapter 13
Teacher Guide for The Ominous Phone Call and Evaluating the Docs

Answer Key

Presentation Skills: The "doctor's" presentation style is …

Student

_____ : <-->

_____ : <-->

_____ : <-->

_____ : <-->

Novice	Competent	Expert
The "doctor" does not act professionally when he/she speaks. The presentation is not revised and the "doctor" seems unprepared to present. Additionally, the "doctor" lacks eye contact and mispronounces many key terms.	The "doctor" could make improvements in one or two areas of presentation; however, these errors do not distract from the presentation itself.	The "doctor" acts professionally when he/she speaks. The presentation contains little to no mistakes. The "doctor" is prepared to present and has great eye contact. He/she has clearly practiced pronouncing key terms.

Written Evaluation: The "doctor's" written assessment is …

Student

_____ : <-->

_____ : <-->

_____ : <-->

_____ : <-->

Novice	Competent	Expert
The "doctor" does not complete the assignment and/or explanations are unclear. There are major flaws in concept mastery and incorrect use of scientific terminology.	The "doctor" completes the assignment but explanations may be slightly ambiguous or unclear. May contain some incompleteness or a cloudy understanding.	The "doctor" shows clarity of thought and assignment fulfills all requirements. The "doctor" shows thorough understanding of scientific content. Statements are supported by evidence.

Answer Key

Chapter 13
Teacher Guide for The Ominous Phone Call and Evaluating the Docs

Final Evaluation: The evaluator will determine if the "doctor" can renew his or her license (which indicates a passing grade) or have his/her license revoked (which indicates more practice in this subject area is needed).

Student

_____: Renew or Revoke

_____: Renew or Revoke

_____: Renew or Revoke

_____: Renew or Revoke

3
AFTER

STUDENT EDITION

Thank you for using this book!

Your opinion is very important.

Please visit
www.StylishSchooling.com
and complete a five-minute teacher survey[1] for *Diagnosis for Classroom Success: Making Anatomy & Physiology Come Alive.*

---

1 To complete the survey, please make sure you have set up a teacher account. Once logged in, the teacher survey can be found under "Contact Us → Teacher Survey." Remember to select "Mystery Diagnoses" from our dropdown menu.

Index

*Page numbers printed in **boldface** type refer to tables or figures.*

A

A Framework for K–12 Science Education: Practices, Crosscutting Concepts, and Core Ideas, 6, 10
Abortion by minor, parental involvement and, 113
Alveoli, 104
Arsenault, A., 5
Arteries, 83, 121
Assessment. *See* Evaluating doctors' performance
Asthma, 104–105

B

Bacterial infections, 81, 87, 94
Birth control, **37**
Blood smears (Lab 3), 33, 79–85
    conclusion of, 84–85
    critical thinking question related to, 84
    lab roles for team members, 80, 81
    after the lesson, 80
    before the lesson, 79–80
    during the lesson, 80
    materials for, 22, 79–80, 82
    medical school notes regarding red and white blood cells, 81
    PowerPoint slide on, 129
    procedure for, 82
    recall questions related to, 83–84
    recording results for each patient, 80, **83**, 84
    role-play comparison chart for, 26
Body mass index (BMI), 73
    chart for adults, 67, **74**
    ranges for, **74**
    recording value for each patient, **75**
    relation to health status, 76
    significance of high BMI, 75–76
Body systems, 133
Bruner, J., 3

C

Carbohydrates, simple and complex, 75
Careers in science, 5–6
Circulatory system, 79, 83, 89, 115, 121–122, 132
Classroom simulations, 4
    comparison chart for, 24–27
Classroom use of lessons, 6
Code of ethics, 39
Common Core State Standards, curriculum alignment with, 4, 15–17
Core Curriculum State Standards (CCSS), 6, 7
Council of Chief State School Offices (CCSSO), 15
Cowan, B. M., 5
Craciun, D., 4
Cultural relevance and pop culture, 5
Curriculum alignment with state standards, 6, 7, 9–17

Index

Common Core State Standards, 15–17
Next Generation Science Standards, 6, 7, 9–14

D
Diabetes (patient #1), 31, 34
   glucose tolerance test for, 11, 16, 25, 27, 108, 109, 111, **111, 112,** 114
   information collected on causes, symptoms, and treatments for, **38**
   lung capacity and breathing difficulty in, 99, 104
   medical record for patient with, 46–47, **54**
   PowerPoint slides on, 129–130
   study group assignment for, 34
   type 1 and type 2, **39,** 107, 113
Diagnosis, 33
   evaluating doctors' performance, 6–7, 33–34, 125–137
   final, 127
   mystery diagnosis rubric, 134–137
   PowerPoint slides on, 129–131
Diagnostic tests, 33
   blood smears (Lab 3), 33, 79–85
   digestive by-products and body mass index analysis (Lab 2), 67–77
   HIV test (Lab 4), 33, 87–95
   hormone test (Lab 6), 33, 107–114
   lung capacity (Lab 5), 33, 97–105
   materials for, 19–23
   urinalysis (Lab 1), 59–65
Dialysis, 64
Digestive by-products and body mass index analysis (Lab 2), 33, 67–77
   body mass index, 73
      chart for adults, **74**
      ranges for, **74**
      recording for each patient, **75**
   chemical observations for, 71–72
   conclusion of, 76–77
   critical thinking question related to, 76
   lab roles for team members, 68–69, 70
   after the lesson, 69
   before the lesson, 67–68
   during the lesson, 68–69
   materials for, 21, 68, 70–71
   meaning of nutrient indicator results, **73**
   nutrition content found in patients' digestive by-products, **72**
   physical observations for, 71
   PowerPoint slide on, 128
   procedure for, 71–73
   recall questions related to, 75–76
   recording results for each patient, 69, 77
   role-play comparison chart for, 25
   setup for, 71
Digestive system, 67, 69, 75, 115, 118–119, 132
Dissection lab. *See* Organ donation surgery

E
Earning your white coats, 31–40
   Hippocratic oath, 39–40
      discussion questions related to, **41**
   information collected on diabetes, **38**
   information collected on HIV infection, **36**
   information collected on pregnancy, **37**
   information collected on sickle cell anemia, **35**
   after the lesson, 32
   before the lesson, 31
   during the lesson, 31–32
   medical school attendance and graduation, 32, 34, 39
   resources for, 31
   study group assignments, 34
   study group at the library, 34
   task overview, 33–34
Endocrine system, 107, 112, 115, 116, 132
Evaluating doctors' performance, 6–7, 33–34, 125–137
   filling out a prescription, 127
   after the lesson, 126
   before the lesson, 125–126
   during the lesson, 126
   mystery diagnosis rubric for, 134–137
   oral/visual assessment options for, 125, 128–132
   recruiting an authentic audience for, 126
   resources for, 125
   written assessment options for, 125, 132–133
Excretory system, 59, 61, 64, 69, 115, 119–120, 132

Index

Exhalation, 103

F
Feedback, 17, 34
Fertilization, **37**
Filling out a prescription, 127
Fungal infections, 81

G
Genetic disorders, 81, 133
Glucose tolerance test, 11, 16, 25, 27, 108, 109, 114
    graphing results of each patient's test, **112**
    medical school notes regarding normal and diabetic test results, **111**
    procedure for, 111
    recording results of each patient's test, **112**
Glycemic index (GI), 67, 75, 107
Glycogen, 107, 113

H
Hankes, J., 5
Hemoglobin protein on red blood cells, in sickle cell anemia, 26, **35, 81,** 84, 99, 104
Hippocratic oath, 32, 33, 39, 45, 64
    discussion questions related to, **41**
    modernized version of, 39–40
Homeostasis, 14, 59, 113, 122, 133
Hormone test (Lab 6), 33, 107–114
    conclusion of, 113–114
    critical thinking question related to, 113
    lab roles for team members, 108, 110
    after the lesson, 108
    before the lesson, 107–108
    during the lesson, 108
    materials for, 23, 108, 110
    medical school notes regarding hCG and insulin, **109**
    PowerPoint slide on, 129
    procedure for glucose tolerance test, 111
        graphing results of each patient's test, **112**
        medical school notes regarding normal and diabetic test results, **111**
        recording results of each patient's test, **112**
    procedure for pregnancy test, 110
        meaning of hCG indicator results, **111**
        recording results for each patient, **111**
    recall questions related to, 112–113
    role-play comparison chart for, 27
Howard, T. C., 5
Human chorionic gonadotropin (hCG), 27, **37,** 107, 108, **109,** 110, 112, 114
    meaning of hCG indicator results, **111**
    recording results of each patient's pregnancy test, **111**
Human immunodeficiency virus (HIV) infection (patient #3), 31, 34, 87
    information collected on causes, symptoms, and treatments for, **36**
    lung capacity and breathing difficulty in, 99–100, 104
    medical record for patient with, 50–51, **56**
    modes of transmission of, **36**
    PowerPoint slides on, 130
    preventing transmission of, 93
    progression to AIDS, 93
    study group assignment for, 34
    susceptibility to other infections in, 93, 99–100
Human immunodeficiency virus (HIV) test (Lab 4), 33, 87–95
    conclusion of, 94–95
    critical thinking question related to, 94
    lab roles of team members, 89, 90
    after the lesson, 89
    before the lesson, 87–88
    during the lesson, 89
    materials for, 22, 88, 90–91
    PowerPoint slide on, 129
    procedure for, 91–93
    recall questions related to, 93–94
    recording results for each patient, 89, **93,** 95
    results from limited sex partner demonstration, **92**
    results from multiple sex partner demonstration, **92**
    role-play comparison chart for, 26
    time after initial infection before HIV antibodies show up on, 94
Hypothesis generation, 10, 33, 44, 45

Index

I
Immune system, 87, 89, 93, 115, 132
Infectious diseases, 81, 87, 94
Inhalation, 103–104
Insulin, 27, **38,** 67, 107, 109, **109, 112,** 113, 114

K
Kane, J. M., 5
Kidney dialysis, 64
Kidney function, 64, 119–120
Kids Health website, 31
Kirkland, K., 5

L
Ladson-Billings, G., 4
Lessons
   alignment with state standards, 6, 7, 9–17
   classroom use of, 6
   cultural relevancy of, 5
   materials for labs, 6, 7, 19–23 (See also specific labs)
   narrative approach of, 3–4
   pacing chart for, 24
   support for implementation of, 7
   use of role-play in, 4
      comparison chart for, 24–27
   use of Teacher Edition for, 7
Lifestyle choices, 133
Lung capacity (Lab 5), 33, 97–105
   acceptable value for height, weight, and gender, 102, **103**
      each patient's actual lung capacity compared to, **103**
   calculations for determination of, **101**
   conclusion of, 105
   critical thinking question related to, 104–105
   determining by balloon's diameter, **102**
   factors that may cause increase or decrease in, **99**
   lab roles for team members, 98, 100
   after the lesson, 98
   before the lesson, 97
   during the lesson, 98
   materials for, 23, 97, 100
   PowerPoint slide on, 129
   procedure for, 100–102
   recall questions related to, 103–104
   recording results for each patient, 98, 105
   role-play comparison chart for, 27

M
Materials for labs, 6, 7, 19–23. See also specific labs
Mayo Clinic website, 31
Medical records, analysis of, 43–53, **54–57**
   after the lesson, 44
   before the lesson, 43
   during the lesson, 43–44
   for patient #1 (diabetes), 46–47, **54**
   for patient #2 (pregnancy), 48–49, **55**
   for patient #3 (HIV), 50–51, **56**
   for patient #4 (sickle cell anemia), 52–53, **57**
   task overview for, 45
Medical school attendance and graduation, 32, 34, 39
Medicare, 64
Milne, C., 3, 5
Musculoskeletal system, 132
Mystery diagnosis rubric, 134–137

N
Narratives as teaching tool, 3–4
National Governors Association Center for Best Practices (NGA Center), 15
National Science Foundation's Project Synthesis, 6
Nervous system, 132
Next Generation Science Standards (NGSS), lesson alignment with, 6, 7, 9–14
   dimension I: practices, 10–12
   dimension II: crosscutting concepts, 12–14
   dimension III: disciplinary core ideas, 14
Nutrition terms, 67
Nutritional guidelines for school meals, 76

O
Obesity and overweight, 73, **74, 75,** 76
Oral/visual assessment options, 125, 128–132
   PowerPoint presentation, 128–131
   public service announcement, 131–132
Organ donation surgery (Lab 7), 33, 115–123

conclusion of, 123
critical thinking question related to, 122
lab roles for team members, 116, 117
    after the lesson, 116
    before the lesson, 115
    during the lesson, 116
materials for, 23, 115–116, 117
procedure for, 118–122
    organ preservation, 122
    preparing for rat dissection, 118, **118**
recall questions related to, 122
Organ donor programs, 122
Osmosis Jones, 87
Outdoor air quality, 105

P
Pacing chart for lessons, 24
Pappas, C. C., 5
Pathogens, 81
Pop culture and cultural relevancy, 5
PowerPoint presentation, 128–131
Pregnancy (patient #2), 31, 34
    information collected on causes, symptoms, and treatments in, **37**
    lung capacity and breathing difficulty in, 99, 104
    medical record for patient during, 48–49, **55**
    PowerPoint slides on, 130
    study group assignment for, 34
Pregnancy test, 27, 108
    meaning of hCG indicator results, **111**
    procedure for, 110
    recording results for each patient, **111**
Privacy of patients, 40, 45
Prognosis, 33, 127
Pub Med Health website, 31
Public service announcement (PSA), 131–132

R
Red blood cells (RBCs), 81, **81, 83**
    in sickle cell anemia, **35, 81,** 99
Reproductive system, 107, 112, 115, 116, 120, 132
Research task, 33–34
    evaluating doctors' performance on, 6–7, 33–34, 125–137

study group assignments for, 34, **35–38**
Residency programs, 32, 45
Respiratory system, 97, 103, 115, 116, 132
Role-play, 4
    comparison chart for, 24–27
Rubric for evaluating doctors' performance, 134–137

S
Safe sex, 90
School meals, nutritional guidelines for, 76
Science education
    cultural relevancy of, 5
    curriculum alignment with state standards for, 6, 7, 9–17
    four needs in, 5–6
    use of role-play in, 4
        comparison chart for, 24–27
    value of narratives in, 3–4
Sexually transmitted diseases (STDs), 87, 90
Sickle cell anemia (patient #4), 31, 34, 84
    hemoglobin protein on red blood cells in, 26, **35, 81,** 84, 99, 104
    information collected on causes, symptoms, and treatments for, **35**
    lung capacity and breathing difficulty in, 99, 104
    medical record for patient with, 52–53, **57**
    newborn screening for, 26
    PowerPoint slides on, 131
    study group assignment for, 34
Sickle cell trait, 84
    sports participation at high altitude and, 84
Spirometer, 99
State standards, curriculum alignment with, 6, 7, 9–17
    Common Core State Standards, 15–17
    Next Generation Science Standards, 6, 7, 9–14
Storytelling as teaching tool, 3–4
Study groups in library, 34
    assignment of, 34
Stylish Schooling, 7
Super Size Me, 67, 79
Surgery. *See* Organ donation surgery
Systems of the body, 133

Index

T
Tate, W., 4
Treatments, 127
    for diabetes, **38**
    for HIV infection, **36**
    during pregnancy, **37**
    for sickle cell anemia, **35**

U
Urinalysis (Lab 1), 33, 59–65
    chemical observations for, 62–63
    conclusion of, 64
    critical thinking question related to, 64
    lab roles for team members, 60
    after the lesson, 60
    before the lesson, 59
    during the lesson, 60
    materials for, 20, 59–60, 61–62
    meaning of nutrient indicator test results, **63**
    medical school notes regarding results of, **61**
    physical observations for, 62
    PowerPoint slide on, 128
    procedure for, 62–63
    recall questions related to, 64
    recording results for each patient, 60, **63,** 65
    role-play comparison chart for, 25
    setup for, 62

V
Vaccines, 94
Varelas, M., 5
Veins, 83, 121
Viral infections, 81, 87, 94

W
White blood cells (WBCs), 81, **81, 83,** 87, 93
    in HIV infection, **81,** 90
Written assessment options, 125, 132–133
    body systems, 133
    how genes and lifestyle choices affect homeostasis, 133

DIAGNOSIS FOR CLASSROOM SUCCESS

Making Anatomy + Physiology Come Alive

Student Edition

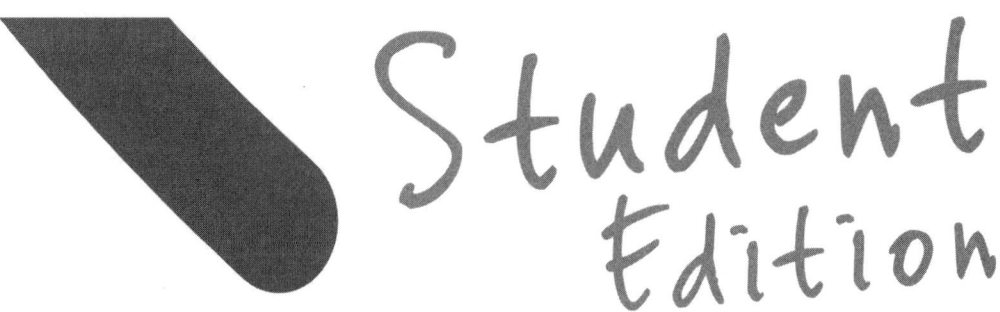

Nicole H. Maller

DIAGNOSIS FOR CLASSROOM SUCCESS

Making Anatomy + Physiology Come Alive

Student Edition

National Science Teachers Association

Arlington, Virginia

National Science Teachers Association

Claire Reinburg, Director
Jennifer Horak, Managing Editor
Andrew Cooke, Senior Editor
Wendy Rubin, Associate Editor
Amy America, Book Acquisitions Coordinator

ART AND DESIGN
Will Thomas Jr., Director
Joe Butera, Senior Graphic Designer, cover and interior design
Images courtesy of ThinkStock.

PRINTING AND PRODUCTION
Catherine Lorrain, Director

NATIONAL SCIENCE TEACHERS ASSOCIATION
David L. Evans, Executive Director
David Beacom, Publisher
1840 Wilson Blvd., Arlington, VA 22201
www.nsta.org/store
For customer service inquiries, please call 800-277-5300.

Copyright © 2013 by the National Science Teachers Association.
All rights reserved. Printed in the United States of America.
16 15 14 13   4 3 2 1

NSTA is committed to publishing material that promotes the best in inquiry-based science education. However, conditions of actual use may vary, and the safety procedures and practices described in this book are intended to serve only as a guide. Additional precautionary measures may be required. NSTA and the authors do not warrant or represent that the procedures and practices in this book meet any safety code or standard of federal, state, or local regulations. NSTA and the authors disclaim any liability for personal injury or damage to property arising out of or relating to the use of this book, including any of the recommendations, instructions, or materials contained therein.

PERMISSIONS
Book purchasers may photocopy, print, or e-mail up to five copies of an NSTA book chapter for personal use only; this does not include display or promotional use. Elementary, middle, and high school teachers may reproduce forms, sample documents, and single NSTA book chapters needed for classroom or noncommercial, professional-development use only. E-book buyers may download files to multiple personal devices but are prohibited from posting the files to third-party servers or websites, or from passing files to non-buyers. For additional permission to photocopy or use material electronically from this NSTA Press book, please contact the Copyright Clearance Center (CCC) (www.copyright.com; 978-750-8400). Please access www.nsta.org/permissions for further information about NSTA's rights and permissions policies.

Library of Congress Cataloging-in-Publication Data
Maller, Nicole H.
  Diagnosis for classroom success: making anatomy and physiology come alive/by Nicole H. Maller.—Student edition.
      pages cm
  Includes index.
  ISBN 978-1-936959-50-1
  1. Diagnosis—Study and teaching. 2. Human anatomy—Study and teaching. 3. Human physiology—Study and teaching. 4. Diagnosis. I. Title.
  RC71.3.M273 2013b
  612.0076—dc23
                         2013009444

CONTENTS

About the Author vii

Acknowledgments ix

Chapter 1: Earning Your White Coats:
Medical School Research 1

Chapter 2: What's Wrong With Me, Doc?
Analyzing Medical Records 11

Chapter 3: Let's Diagnose Them, Lab 1
Urinalysis 25

Chapter 4: Let's Diagnose Them, Lab 2
Digestive By-Products and Body Mass Index Analysis 33

Chapter 5: Let's Diagnose Them, Lab 3
Blood Smears 43

Chapter 6: Let's Diagnose Them, Lab 4
HIV Test 49

Chapter 7: Let's Diagnose Them, Lab 5
Lung Capacity 57

Chapter 8: Let's Diagnose Them, Lab 6
Hormone Test 65

Chapter 9: Emergency! Lab 7
Performing Surgery 71

Chapter 10: The Ominous Phone Call 79

Chapter 11: Evaluating the Docs 81

Index 93

About the Author

Nicole H. Maller received a B.S. in Teaching Biology 7–12 from New York University in 2006 and an M.A. in Science Education from New York University in 2010. Her career in education began in Williamsburg, Brooklyn at The Green School: An Academy for Environmental Careers. A year later, she relocated to Manhattan and worked at Vanguard High School, where she continues to teach Living Environment to tenth graders and a Biopsychology course she developed specifically for 11th and 12th graders. During her summers, Nicole teaches Introductory Chemistry and Introductory Forensics at Columbia University's six-week, Upward Bound program to first-generation college-bound students. She also tutors middle school and high school students in Manhattan.

Acknowledgments

Vanguard High School
for providing teachers the freedom to teach students the best way they know how

Catherine Bell
for helping me make this vision come to life in the classroom

NYU Professors
Dr. Pamela Fraser-Abder, Catherine Milne, Jason Blonstein, and Bob Wallace
for your guidance and professional insight

Tal Savariego
for your continuous support and editing skills

Jaimie Glick, M.D.
for evaluating my 'Docs' at round tables

Adam Handler, M.D.
for providing feedback and editing for medical accuracy

Family and Friends
for listening to and believing in my ideas

Chapter 1

Earning Your White Coats
Medical School Research

Task Overview

To successfully complete the Anatomy and Physiology unit, you and your classmates will be required to:

1. *Attend and graduate medical school*: In order to graduate medical school, all students must complete the research portion of this project. All medical school students will be required to investigate, as thoroughly as possible, the causes of, symptoms of, and potential treatments for four health conditions. Since there will be limited time to complete this task, working efficiently as a group will be critical. Once the research portion is approved, permission to graduate will be granted by the medical school president (your teacher).

2. *Sign the Hippocratic oath*: Graduating medical students will be required to read and sign the Hippocratic oath before accepting and treating patients, ensuring all soon-to-be doctors understand the role of ethics in medicine.

3. *Meet your patients*: Based on both the knowledge obtained from medical school and the medical records provided by the four patients, doctors (you the students) will develop an initial hypothesis.

4. *Run diagnostic tests on patients*: Doctors will conduct six labs to help diagnose the four patients. Lab 7 will not assist in the diagnosis of *your* four primary patients.

   - Lab 1: Urinalysis
   - Lab 2: Digestive By-Product and BMI Analysis
   - Lab 3: Blood Smears
   - Lab 4: HIV Test
   - Lab 5: Lung Capacity

Chapter 1
Earning Your White Coats: Medical School Research

- Lab 6: Hormone Test
- Lab 7: Performing Surgery

5. *Diagnose patients and develop a prognosis*: Once the group reaches a consensus regarding each patient's appropriate diagnosis, the medical chart must be completed so that (a) all patients can thoroughly understand their prognosis and (b) a prescribed treatment can be filled by a pharmacist.

6. *Develop a written, visual, and/or oral report*: All doctors will be evaluated on their ability to collect and analyze *evidence*, their ability to make *connections* between the biology content and the various laboratories used to diagnose patients, and on their understanding of the topics discussed.

7. *Receive feedback from your evaluators*. Evaluators will determine whether or not a doctor may continue practicing medicine (and has therefore passed) or if a doctor is at risk of losing his or her license (and is therefore not familiar enough with the content).

Welcome to Medical School!

Greetings! Your professor has assigned you and a team of medical students to conduct research on the following four health conditions: sickle cell anemia, the human immunodeficiency virus (HIV), pregnancy, and diabetes. Within your team's research, be sure to include the causes of, symptoms of, and treatments of (if any exist) the aforementioned conditions. Your professor (classroom teacher) will determine the time allotted to complete this task. Remember, medical school requires dedication, hard work, and great attention to detail. Stay focused and good luck!

Study Group at the Library

You and your team have headed straight to the medical school library. As a group, you decide it is best to split up research tasks and share your findings afterward. Before starting, each group member selects one of the four health conditions to study. Everyone in the group promises to complete the table (Tables 1.1, 1.2, 1.3, and 1.4) that corresponds with the assigned condition. When everyone has finished, be sure to communicate your findings to others so that they, too, learn about the condition.

Chapter 1
Earning Your White Coats: Medical School Research

Study Group Assignments (Student Name)

1. _____ will study sickle cell anemia and complete Table 1.1.
2. _____ will study human immunodeficiency virus and complete Table 1.2.
3. _____ will study pregnancy and complete Table 1.3.
4. _____ will study diabetes and complete Table 1.4.

Chapter 1
Earning Your White Coats: Medical School Research

TABLE 1.1. INFORMATION COLLECTED ON SICKLE CELL ANEMIA

Sickle Cell Anemia	
Causes 1. How does one get sickle cell anemia? 2. How does one get the sickle cell trait?	
Symptoms 1. What are the symptoms of sickle cell anemia? 2. What cells are affected by sickle cell? 3. What shape do these cells turn into? 4. What protein is mutated on these cells? Explain how this is related to symptoms of sickle cell anemia.	
Treatment/cures 1. What cures exist? 2. What treatments exist?	

Chapter 1
Earning Your White Coats: Medical School Research

TABLE 1.2. INFORMATION COLLECTED ON HUMAN IMMUNODEFICIENCY VIRUS (HIV)

Human Immunodeficiency Virus (HIV)	
Causes 1. How is HIV contracted? 2. Is HIV caused by a virus or bacteria? 3. What type of cell does HIV attack in the immune system?	
Symptoms 1. List the symptoms at different stages of an HIV infection.	(a) Directly after infection: (b) 3–6 months after infection: (c) Years after the infection:
Treatment/cures 1. What cures exist? 2. What treatments exist?	

DIAGNOSIS FOR CLASSROOM SUCCESS: Making Anatomy ● Physiology Come Alive

TABLE 1.3. INFORMATION COLLECTED ON PREGNANCY

Pregnancy	
Causes 1. What is fertilization? 2. How does fertilization take place? 3. Where in the female does the egg get fertilized? 4. Where does the female egg travel upon fertilization?	
Symptoms 1. What are the symptoms of pregnancy?	(a) What hormone is released by the placenta and detected by a pregnancy test? What are the symptoms of the (b) first trimester? (c) second trimester? (d) third trimester?
Treatment/cures 1. What methods of birth control exist? Describe them. 2. What preventative health measures are recommended for an expecting mother?	

Chapter 1
Earning Your White Coats: Medical School Research

TABLE 1.4. INFORMATION COLLECTED ON DIABETES

Diabetes	
Causes 1. What is diabetes? 2. What organ does not function properly in a diabetic? 3. What is insulin? 4. How does someone obtain type 1 diabetes? 5. How might someone develop type 2 diabetes?	
Symptoms 1. What symptoms will a diabetic experience?	
Treatment/cures 1. How does a person manage type 1 diabetes? 2. How does a person manage type 2 diabetes?	

Chapter 1
Earning Your White Coats: Medical School Research

Medical School President's Signature of Approval:

_____

Graduation

Congratulations! You may *officially* graduate medical school! As a graduating medical school class, you must read and sign the modernized version of the Hippocratic oath. According to the Public Broadcasting Station's website, the Hippocratic oath originated in fifth century BC as a means of protecting patients with a code of ethics to be followed by health care professionals and physicians. Due to the outdated nature of the original Hippocratic oath, with its references to gods, goddesses, and slavery, it was revised in 1964.

As you read the modernized version of the Hippocratic oath aloud, annotate the text. Place a question mark next to sentences that puzzle you, a "W" next to sentences that worry or concern you, and an "E" next to sentences that excite or seem beneficial to you.

Chapter 1
Earning Your White Coats: Medical School Research

The Hippocratic Oath (The Modern Version)

I swear to fulfill, to the best of my ability and judgment, this covenant:

I will respect the hard-won scientific gains of those physicians in whose steps I walk, and gladly share such knowledge as is mine with those who are to follow.

I will apply, for the benefit of the sick, all measures [that] are required, avoiding those twin traps of over treatment and therapeutic nihilism.

I will remember that there is art to medicine as well as science, and that warmth, sympathy, and understanding may outweigh the surgeon's knife or the chemist's drug.

I will not be ashamed to say "I know not," nor will I fail to call in my colleagues when the skills of another are needed for a patient's recovery.

I will respect the privacy of my patients, for their problems are not disclosed to me that the world may know. Most especially must I tread with care in matters of life and death.

If it is given me to save a life, all thanks. But it may also be within my power to take a life; this awesome responsibility must be faced with great humbleness and awareness of my own frailty. Above all, I must not play at God.

I will remember that I do not treat a fever chart, a cancerous growth, but a sick human being, whose illness may affect the person's family and economic stability. My responsibility includes these related problems, if I am to care adequately for the sick.

I will prevent disease whenever I can, for prevention is preferable to cure.

I will remember that I remain a member of society, with special obligations to all my fellow human beings, those sound of mind and body as well as the infirm.

If I do not violate this oath, may I enjoy life and art, respected while I live and remembered with affection thereafter. May I always act so as to preserve the finest traditions of my calling and may I long experience the joy of healing those who seek my help.

—Written in 1964 by Louis Lasagna, Academic Dean of the School of Medicine at Tufts University, and used in many medical schools today.

Your Signature: _____

TABLE 1.5. HIPPOCRATIC OATH DISCUSSION QUESTIONS

As you read through the Hippocratic oath, which statements or terms puzzled you?	
What concerns you about the Hippocratic oath? (Cons)	What seems promising about the Hipocratic oath? (Pros)
What is your stance on the Hippocratic oath?	
What suggestions might you recommend if the Hippocratic oath were revised again?	

Chapter 2
What's Wrong With Me, Doc?
Analyzing Medical Records

Task Overview

Congratulations! After applying to several residency programs, you finally found a job at Vanguard Hospital. Vanguard Hospital requires that you collaborate with your colleagues when treating patients. Your first day at work is extremely busy! Four celebrity patients have been rushed into your hospital with very alarming symptoms. Using your knowledge from medical school regarding the risk factors and symptoms of sickle cell anemia, HIV, pregnancy, and diabetes, hypothesize the condition afflicting each patient.

Medical Records

On the following eight pages are your four patients' medical records. Included in these records are their symptoms, lifestyle habits, and family medical histories. As a team, analyze these medical records. Remember, these medical records contain confidential information and the Hippocratic oath signed in medical school states that you "will respect the privacy of [your] patients, for their problems are not disclosed to [you, so] that the world may know." Therefore, it remains essential that you discuss your patients' health solely with your team. When your team has developed a hypothesis for each patient, record it on his or her designated medical chart under the section Medical Record. Medical charts for Patients #1, #2, #3 and #4 can be found in Tables 2.1, 2.2, 2.3, and 2.4 (pp. 20–23) respectively. Be sure to cite evidence as to why you believe each patient has the health condition you suggest. These medical charts will play a vital role in your final diagnosis, so be as detailed as possible.

Chapter 2
What's Wrong With Me, Doc? Analyzing Medical Records

VANGUARD HOSPITAL
Medical Record

Patient's Name: <u>Jane Smith</u>
Date of Birth: <u>3/18/1954</u> Sex: <u>F</u> Height: <u>5'6''</u> Weight: <u>200 lbs</u>

Why are you here? <u>Had a dizzy spell and fainted. My vision seems blurry.</u>

Current Medications (prescription and non-prescription, vitamins, home remedies, birth control pills, herbs): <u>None!</u>

Personal Medical History (please indicate whether you have had any of the following medical problems):

___Congenital Heart Disease ___Depression/suicide attempt
___Myocardial Infarction (heart attack) ___Alcoholism
<u>X</u> Hypertension (high blood pressure) ___If you ever had a blood transfusion
___Diabetes (trouble regulating blood sugar) ___Abnormal pap smear (at gynecologist)
___High cholesterol (fat in blood) ___Other problems (specify) _____
___Stroke (clogged artery to brain) ___Cancer
___Coagulation (bleeding/clotting) disorder <u>X</u> Thyroid problem

Women's Gynecological History:
# of pregnancies <u>1</u> # of deliveries <u>0</u> # of abortions <u>1</u> # of miscarriages <u>0</u>
Do you have any concerns about your periods? <u>No</u>
Do you have any concerns about menopause? <u>No</u>

Family History: Indicate with a check mark which family members have had any of the following:

Condition	Mom	Dad	Sis	Bro	Other relative	Condition	Mom	Dad	Sis	Bro	Other relative
Alcoholism						Hearing problems					
Anemia						Heart Attack (coronary artery disease)		X			
Arthritis (joint problems)						Hypertension (high blood pressure)	X	X			
Asthma						High cholesterol		X			
Bleeding Problems						Kidney disease					
Cancer	X					Migraine headaches					
Depression						Osteoporosis (weak bones)					
Diabetes, Type 1 (childhood onset)						Stroke		X			
Diabetes, Type II (adult onset)			X	X		Thyroid disorders	X			X	
Eczema (itchy skin)						Tuberculosis					
Epilepsy (seizures)						Neural disorders					
Genetic diseases											
Glaucoma (vision problem)											

Chapter 2
What's Wrong With Me, Doc? Analyzing Medical Records

Social History:

Tobacco use: ~~Quit:~~ Date _____   Alcohol Use
    (Never)   Do you drink? yes
    ~~Current smoker:~~ packs/day ____ # of yrs ____   If yes, #drinks/wk 1/wk

Do you use any recreational drugs? no
Have you ever used needles? no
Do you exercise regularly? no

Sexuality:
Are you sexually active? yes   Current sex partner(s) is/are: (Male) ~~Female~~
Birth control method: I'm on birth control pills.
Do you practice safe sex? We are in a committed relationship, so not always.
Have you ever had a sexually transmitted disease (STDs)? no   If yes, please list: _____
Are you interested in being screened for a sexually transmitted disease? Since I'm here, yes.

Current Symptoms:

Constitutional
___ Fevers/chills/sweats
X Unexplained weight loss/gain
X Fatigue Weakness
X Excessive thirst or urination
Eyes
X Change of vision
Ears/Nose/Throat/Mouth
___ Difficult hearing/ringing in ears
___ Problems with teeth/gums
___ Allergies
Cardiovascular
___ Chest pain/discomfort
___ Leg pain with exercise
___ Palpitations
Chest (breast)
___ Lump or discharge
Respiratory
___ Cough/Wheeze
X Difficulty breathing
Gastrointestinal (digestive)
___ Abdominal pain
___ Blood in bowl movement
___ Nausea/vomiting/diarrhea

Genitourinary
___ Nighttime urination
___ leaking urine
___ Unusual vaginal bleeding
___ Discharge: penis or vagina
___ Sexual function problems
Musculo-skeletal
___ Muscle/joint pain
Skin
___ Rash or mole change
Neurological
X Headaches
X Dizziness/light-headedness
X Numbness
___ Memory loss
___ Loss of coordination
Psychiatric
___ Anxiety/stress
X Problems with sleep
___ Depression
Blood/Lymphatic (immune)
___ Unexplained lumps
___ Easy bruising/bleeding
Other: _____

Socio-economic:
Occupation: Actress/Host/Editor
Education completed: High School
Marital Status: Not married
Children: 0
Who lives at home with you: My boyfriend

Chapter 2
What's Wrong With Me, Doc? Analyzing Medical Records

VANGUARD HOSPITAL
Medical Record

Patient's Name: Cindy Jones
Date of Birth: 12/8/69   Sex: F   Height: 5'2''   Weight: 120 lbs

Why are you here? Been feeling nauseous.

Current Medications (prescription and non-prescription, vitamins, home remedies, birth control pills, herbs): birth control pills

Personal Medical History (please indicate whether you have had any of the following medical problems):
- ___Congenital Heart Disease
- ___Myocardial Infarction (heart attack)
- ___Hypertension (high blood pressure)
- ___Diabetes (trouble regulating blood sugar)
- ___High cholesterol (fat in blood)
- ___Stroke (clogged artery to brain)
- ___Coagulation (bleeding/clotting) disorder
- ___Depression/suicide attempt
- ___Alcoholism
- ___If you ever had a blood transfusion
- _X_Abnormal pap smear (at gynecologist)
- ___Other problems (specify) _____
- ___Cancer
- ___Thyroid problem

Women's Gynecological History:
# of pregnancies 3   # of deliveries 2   # of abortions 1   # of miscarriages 0
Do you have any concerns about your periods? Yes
Do you have any concerns about menopause? No

Family History: Indicate with a check mark which family members have had any of the following:

Condition	Mom	Dad	Sis	Bro	Other relative	Condition	Mom	Dad	Sis	Bro	Other relative
Alcoholism						Hearing problems					
Anemia	X		X			Heart Attack (coronary artery disease)					
Arthritis (joint problems)						Hypertension (high blood pressure)					
Asthma						High cholesterol					
Bleeding Problems						Kidney disease					
Cancer	X					Migraine headaches	X		X		
Depression		X	X			Osteoporosis (weak bones)					
Diabetes, Type 1 (childhood onset)						Stroke					
Diabetes, Type II (adult onset)						Thyroid disorders					
Eczema (itchy skin)						Tuberculosis					
Epilepsy (seizures)						Neural disorders					
Genetic diseases											
Glaucoma (vision problem)											

Chapter 2
What's Wrong With Me, Doc? Analyzing Medical Records

Social History:

Tobacco use: Quit: Date <u>2 years ago ☺</u>   Alcohol Use:
    Never   Do you drink? <u>yes</u>
    Current smoker: packs/day ____ # of yrs ____   If yes, #drinks/wk <u>3/wk</u>

Do you use any recreational drugs? <u>no</u>
Have you ever used needles? <u>no</u>
Do you exercise regularly? <u>yes</u>

Sexuality:
Are you sexually active? <u>yes</u>   Current sex partner(s) is/are: (Male) ~~Female~~
Birth control method: <u>I'm on birth control pills.</u>
Do you practice safe sex? <u>Since I'm on the pill, I don't always use condoms.</u>
Have you ever had a sexually transmitted disease (STDs)? <u>yes</u>   If yes, please list: <u>H.P.V.</u>
Are you interested in being screened for a sexually transmitted disease? <u>yes</u>

Current Symptoms:

Constitutional
___ Fevers/chills/sweats
X Unexplained weight loss/gain
X Fatigue Weakness
X Excessive thirst or urination
Eyes
___ Change of vision
Ears/Nose/Throat/Mouth
___ Difficult hearing/ringing in ears
___ Problems with teeth/gums
___ Allergies
Cardiovascular
___ Chest pain/discomfort
___ Leg pain with exercise
___ Palpitations
Chest (breast)
___ Lump or discharge
Respiratory
___ Cough/Wheeze
X Difficulty breathing
Gastrointestinal (digestive)
<u>Ab</u>dominal pain
___ Blood in bowel movement
<u>N</u>ausea/vomiting/diarrhea

Genitourinary
___ Nighttime urination
___ Leaking urine
___ Unusual vaginal bleeding
___ Discharge: penis or vagina
___ Sexual function problems
Muscular-skeletal
___ Muscle/joint pain
Skin
___ Rash or mole change
Neurological
X Headaches
X Dizziness/light-headedness
___ Numbness
___ Memory loss
___ Loss of coordination
Psychiatric
___ Anxiety/stress
___ Problems with sleep
___ Depression
Blood/Lymphatic (immune)
___ Unexplained lumps
___ Easy bruising/bleeding
Other: _____

Socio-economic:
Occupation: <u>Singer/Actress</u>
Education completed: <u>High School</u>
Marital Status: <u>Single</u>
Children: <u>2</u>
Who lives at home with you: <u>My two kids</u>

Chapter 2
What's Wrong With Me, Doc? Analyzing Medical Records

VANGUARD HOSPITAL
Medical Record

Patient's Name: <u>John Thomas</u>
Date of Birth: <u>3/1/1989</u>   Sex: <u>M</u>   Height: <u>6'2''</u>   Weight: <u>140 lbs</u>

Why are you here? <u>Strange rash, chills, and fever</u>

Current Medications (prescription and non-prescription, vitamins, home remedies, birth control pills, herbs):
<u>none</u>

Personal Medical History (please indicate whether you have had any of the following medical problems):
___Congenital Heart Disease   ___Depression/suicide attempt
___Myocardial Infarction (heart attack) ___Alcoholism
___ Hypertension (high blood pressure)   ___If you ever had a blood transfusion
___Diabetes (trouble regulating blood sugar) ___Abnormal pap smear (at gynecologist)
___High cholesterol (fat in blood)   ___Other problems (specify) _____
___Stroke (clogged artery to brain)   ___Cancer
___Coagulation (bleeding/clotting) disorder ___ Thyroid problem

Women's Gynecological History:
# of pregnancies ____   # of deliveries ____   # of abortions ____   # of miscarriages ____
Do you have any concerns about your periods? ____
Do you have any concerns about menopause? ____

Family History: Indicate with a check mark which family members have had any of the following:

Condition	Mom	Dad	Sis	Bro	Other relative	Condition	Mom	Dad	Sis	Bro	Other relative
Alcoholism		X				Hearing problems					
Anemia						Heart Attack (coronary artery disease)					
Arthritis (joint problems)						Hypertension (high blood pressure)	X	X			
Asthma						High cholesterol					
Bleeding Problems						Kidney disease					
Cancer:	X					Migraine headaches	X		X		
Depression		X				Osteoporosis (weak bones)					
Diabetes, Type 1 (childhood onset)						Stroke					X
Diabetes, Type II (adult onset)						Thyroid disorders					
Eczema (itchy skin)						Tuberculosis					
Epilepsy (seizures)						Neural disorders					
Genetic diseases											
Glaucoma (vision problem)		X									

Chapter 2
What's Wrong With Me, Doc? Analyzing Medical Records

Social History:

Tobacco use: Quit: Date _____   Alcohol Use:
    Never   Do you drink? yes
    Current smoker: 1 packs/day # of yrs 3 years   If yes, #drinks/wk 3/wk

Do you use any recreational drugs? no
Have you ever used needles? no
Do you exercise regularly? yes

Sexuality:
Are you sexually active? yes   Current sex partner(s) is/are: Male (Female)
Birth control method: condoms
Do you practice safe sex? sometimes
Have you ever had a sexually transmitted disease (STDs)? yes   If yes, please list: gonorrhea
Are you interested in being screened for a sexually transmitted disease? yes

Current Symptoms:

Constitutional
X Fever/chills/sweats
X Unexplained weight loss/gain
X Fatigue Weakness
___ Excessive thirst or urination
Eyes
X Change of vision
Ears/Nose/Throat/Mouth
___ Difficult hearing/ringing in ears
___ Problems with teeth/gums
___ Allergies
Cardiovascular
___ Chest pain/discomfort
___ Leg pain with exercise
___ Palpitations
Chest (breast)
___ Lump or discharge
Respiratory
___ Cough/Wheeze
___ Difficulty breathing
Gastrointestinal (digestive)
___ Abdominal pain
___ Blood in bowl movement
X Nausea/vomiting/diarrhea

Genitourinary
___ Nighttime urination
___ leaking urine
___ Unusual vaginal bleeding
X Discharge: penis or vagina
___ Sexual function problems
Musculo-skeletal
___ Muscle/joint pain
   Skin
X Rash or mole change
Neurological
___ Headaches
X Dizziness/light-headedness
___ Numbness
___ Memory loss
___ Loss of coordination
Psychiatric
___ Anxiety/stress
___ Problems with sleep
___ Depression
Blood/Lymphatic (immune)
X Unexplained lumps
___ Easy bruising/bleeding
Other: _____

Socio-economic:
Occupation: Singer/Dancer
Education completed: High School
Marital Status: Single
Children: Don't think so
Who lives at home with you: My mom, sometimes

Chapter 2
What's Wrong With Me, Doc? Analyzing Medical Records

VANGUARD HOSPITAL
Medical Record

Patient's Name: Robert Smith
Date of Birth: 10/14/82   Sex: M   Height: 5'8''   Weight: 155 lbs

Why are you here?: had some difficulty breathing, been weak after recent concerts

Current Medications (prescription and non-prescription, vitamins, home remedies, birth control pills, herbs): none

Personal Medical History (please indicate whether you have had any of the following medical problems):

- ___ Congenital Heart Disease
- ___ Myocardial Infarction (heart attack)
- ___ Hypertension (high blood pressure)
- ___ Diabetes (trouble regulating blood sugar)
- ___ High cholesterol (fat in blood)
- ___ Stroke (clogged artery to brain)
- ___ Coagulation (bleeding/clotting) disorder
- ___ Depression/suicide attempt
- ___ Alcoholism
- ___ If you ever had a blood transfusion
- ___ Abnormal pap smear (at gynecologist)
- ___ Other problems (specify) _____
- ___ Cancer
- ___ Thyroid problem

Women's Gynecological History:
# of pregnancies ____   # of deliveries ____   # of abortions ____   # of miscarriages ____
Do you have any concerns about your periods? ____
Do you have any concerns about menopause? ____

Family History: Indicate with a check mark which family members have had any of the following:

Condition	Mom	Dad	Sis	Bro	Other relative	Condition	Mom	Dad	Sis	Bro	Other relative
Alcoholism						Hearing problems					X (grandpa)
Anemia	X			X		Heart Attack (coronary artery disease)					
Arthritis (joint problems)						Hypertension (high blood pressure)					
Asthma						High cholesterol					
Bleeding Problems	X		X	X		Kidney disease					
Cancer:						Migraine headaches					
Depression			X			Osteoporosis (weak bones)					
Diabetes, Type 1 (childhood onset)						Stroke					
Diabetes, Type II (adult onset)						Thyroid disorders					
Eczema (itchy skin)						Tuberculosis					
Epilepsy (seizures)						Neural disorders					
Genetic diseases											
Glaucoma (vision problem)		X									

Social History:

Tobacco use: Quit: Date <u>two months ago</u>　　　　Alcohol Use:
　　　Never　　　　　　　　　　　　　　　　　　　　Do you drink? <u>yes</u>
　　　Current smoker: _____ packs/day # of yrs _____　　If yes, #drinks/wk <u>5/wk</u>

Do you use any recreational drugs? <u>Used to</u>
Have you ever used needles? <u>no</u>
Do you exercise regularly? <u>yes</u>

Sexuality:
Are you sexually active? <u>yes</u>　　Current sex partner(s) is/are: Male (Female)
Birth control method: <u>condoms</u>
Do you practice safe sex? <u>every so often</u>
Have you ever had a sexually transmitted disease (STDs)? <u>yes</u>　If yes, please list: <u>Chlamydia</u>
Are you interested in being screened for a sexually transmitted disease? <u>yes</u>

Current Symptoms:

Constitutional
　__ Fevers/chills/sweats
　__ Unexplained weight loss/gain
　X Fatigue Weakness
　__ Excessive thirst or urination
Eyes
　___ Change of vision
Ears/Nose/Throat/Mouth
　___ Difficult hearing/ringing in ears
　___ Problems with teeth/gums
　___ Allergies
Cardiovascular
　X Chest pain/discomfort
　X Leg pain with exercise
　___ Palpitations
Chest (breast)
　___ Lump or discharge
Respiratory
　___ Cough/Wheeze
　X Difficulty breathing
Gastrointestinal (digestive)
　X Abdominal pain
　___ Blood in bowl movement
　___ Nausea/vomiting/diarrhea

Genitourinary
　___ Nighttime urination
　___ Leaking urine
　___ Unusual vaginal bleeding
　___ Discharge: penis or vagina
　___ Sexual function problems
Musculo-skeletal
　X Muscle/joint pain
Skin
　___ Rash or mole change
Neurological
　___ Headaches
　___ Dizziness/light-headedness
　___ Numbness
　___ Memory loss
　___ Loss of coordination
Psychiatric
　___ Anxiety/stress
　___ Problems with sleep
　___ Depression
Blood/Lymphatic (immune)
　___ Unexplained lumps
　___ Easy bruising/bleeding
Other: My eyes and skin seem to have a yellowish color

Socio-economic:
Occupation: <u>Rapper</u>
Education completed: <u>G.E.D.</u>
Marital Status: <u>Single</u>
Children: <u>??????</u>
Who lives at home with you: <u>Self</u>

Chapter 2
What's Wrong With Me, Doc? Analyzing Medical Records

TABLE 2.1. PATIENT #1'S MEDICAL CHART

Vanguard Hospital Medical Chart

Dr. _____ Patient #1: _____

Medical Records
I think they *might* be suffering from _____ because…

-
-
-

Lab 1—Urinalysis
☐ Normal   ☐ Sugar   ☐ Protein   ☐ Bacteria

Lab 2—Digestive By-Products and BMI Analysis
☐ No issues   ☐ Thirst   ☐ Vomit   ☐ Diarrhea   ☐ Constipation
☐ BMI. # is _____   ☐ Healthy   ☐ Underweight   ☐ Overweight   ☐ Obese

Lab 3—Blood Smears
☐ Normal Red Blood Cells   ☐ Sickle Cell
☐ # of Red Blood Cells: _____ million/μL/cu mm
☐ RBC count normal   ☐ RBC too high   ☐ RBC too low
☐ # of White Blood Cells: _____ /μL/cu mm
☐ WBC count normal   ☐ WBC too high   ☐ WBC too low

Lab 4—HIV Test
☐ + test   ☐ – test

Lab 5—Lung Capacity
Lung Volume #: _____ cc
☐ Capacity is normal   ☐ Capacity is too high   ☐ Capacity is too low

Lab 6—Hormone Test
☐ + for hCG   o – for hCG   ☐ No test necessary
☐ Sugar levels above 200 (low insulin)   ☐ Sugar levels between 145 and 200 (borderline)
☐ Sugar levels below 145 (normal insulin)

Diagnosis: _____

Treatments *(research this!)*:
-
-
-

Prognosis *(research this!)*:
-

> Chapter 2
> What's Wrong With Me, Doc? Analyzing Medical Records

TABLE 2.2. PATIENT #2'S MEDICAL CHART

Vanguard Hospital Medical Chart

Dr. _____   Patient #2: _____

Medical Records
I think they *might* be suffering from _____ because...

-
-
-

Lab 1—Urinalysis
☐ Normal   ☐ Sugar   ☐ Protein   ☐ Bacteria

Lab 2—By-Products and BMI Analysis
☐ No issues   ☐ Thirst   ☐ Vomit   ☐ Diarrhea   ☐ Constipation
☐ BMI. # is _____   ☐ Healthy   ☐ Underweight   ☐ Overweight   ☐ Obese

Lab 3—Blood Smears
☐ Normal Red Blood Cells   ☐ Sickle Cell
☐ # of Red Blood Cells: _____ million/µL/cu mm
☐ RBC count normal   ☐ RBC too high   ☐ RBC too low
☐ # of White Blood Cells: _____ /µL/cu mm
☐ WBC count normal   ☐ WBC too high   ☐ WBC too low

Lab 4—HIV Test
☐ + test   ☐ – test

Lab 5—Lung Capacity
Lung Volume #: _____ cc
☐ Capacity is normal   ☐ Capacity is too high   ☐ Capacity is too low

Lab 6—Hormone Test
☐ + for hCG   o – for hCG   ☐ No test necessary
☐ Sugar levels above 200 (low insulin)   ☐ Sugar levels between 145 & 200 (borderline)
☐ Sugar levels below 145 (normal insulin)

Diagnosis: _____

Treatments *(research this!)*:
-
-
-

Prognosis *(research this!)*:
-

TABLE 2.3. PATIENT #3'S MEDICAL CHART

Vanguard Hospital Medical Chart

Dr. _____   Patient #3: _____

Medical Records
I think they *might* be suffering from _____ because…

-
-
-

Lab 1—Urinalysis
☐ Normal   ☐ Sugar   ☐ Protein   ☐ Bacteria

Lab 2—Digestive By-Products and BMI Analysis
☐ No issues   ☐ Thirst   ☐ Vomit   ☐ Diarrhea   ☐ Constipation
☐ BMI. # is _____   ☐ Healthy   ☐ Underweight   ☐ Overweight   ☐ Obese

Lab 3—Blood Smears
☐ Normal Red Blood Cells   ☐ Sickle Cell
☐ # of Red Blood Cells: _____ million/μL/cu mm
☐ RBC count normal   ☐ RBC too high   ☐ RBC too low
☐ # of White Blood Cells: _____ /μL/cu mm
☐ WBC count normal   ☐ WBC too high   ☐ WBC too low

Lab 4—HIV Test
☐ + test   ☐ – test

Lab 5—Lung Capacity
Lung Volume #: _____ cc
☐ Capacity is normal   ☐ Capacity is too high   ☐ Capacity is too low

Lab 6—Hormone Test
☐ + for hCG   o – for hCG   ☐ No test necessary
☐ Sugar levels above 200 (low insulin)   ☐ Sugar levels between 145 & 200 (borderline)
☐ Sugar levels below 145 (normal insulin)

Diagnosis: _____

Treatments *(research this!)*:

-
-
-

Prognosis *(research this!)*:

-

Chapter 2
What's Wrong With Me, Doc? Analyzing Medical Records

TABLE 2.4. PATIENT #4'S MEDICAL CHART

Vanguard Hospital Medical Chart

Dr. _____   Patient #4: _____

Medical Records
I think they *might* be suffering from _____ because…

-
-
-

Lab 1—Urinalysis
☐ Normal   ☐ Sugar   ☐ Protein   ☐ Bacteria

Lab 2—Digestive By-Products and BMI Analysis
☐ No issues   ☐ Thirst   ☐ Vomit   ☐ Diarrhea   ☐ Constipation
☐ BMI. # is _____   ☐ Healthy   ☐ Underweight   ☐ Overweight   ☐ Obese

Lab 3—Blood Smears
☐ Normal Red Blood Cells   ☐ Sickle Cell
☐ # of Red Blood Cells: _____ million/µL/cu mm
☐ RBC count normal   ☐ RBC too high   ☐ RBC too low
☐ # of White Blood Cells: _____ /µL/cu mm
☐ WBC count normal   ☐ WBC too high   ☐ WBC too low

Lab 4—HIV Test
☐ + test   ☐ – test

Lab 5—Lung Capacity
Lung Volume #: _____ cc
☐ Capacity is normal   ☐ Capacity is too high   ☐ Capacity is too low

Lab 6—Hormone Test
☐ + for hCG o – for hCG   ☐ No test necessary
☐ Sugar levels above 200 (low insulin)   ☐ Sugar levels between 145 & 200 (borderline)
☐ Sugar levels below 145 (normal insulin)

Diagnosis: _____

Treatments *(research this!)*:

-
-
-

Prognosis *(research this!)*:

-

Chapter 3
Let's Diagnose Them, Lab 1
Urinalysis

In medical school, you learned about the excretory system and its urinary track. Following proper procedure, you have asked each of your patients to submit urine samples. Nurses have informed you that both patient #1 and patient #2 are experiencing an abnormal increase in urination. Thankfully, a urinalysis can determine the state of one's health by examining physical and chemical properties of urine. Use the next few minutes to review your medical school notes (Table 3.1, p. 26) regarding the interpretation of urinalysis results. Once completed, examine your patients' urine samples.

Chapter 3
Let's Diagnose Them, Lab 1: Urinalysis

Lab Roles (Fill in Names of Team Members)

1. _____ is the task manager (reads procedure and ensures everyone is following proper protocol).

2. _____ is the materials manager (retrieves and returns materials; cleans materials and table).

3. _____ are the doctors (completes lab work, such as adding chemicals, heating chemicals, and so on; to be completed by more than one group member).

4. _____ is the recorder (ensures the group's data is properly recorded).

TABLE 3.1. MEDICAL SCHOOL NOTES REGARDING URINALYSIS RESULTS

Color	If the urine is …	What it could indicate is …
	Dark yellow	• dehydration or fever
	Pale light yellow	• patient drank a lot of liquids prior • diabetes
	Red with blood	• damage to kidneys
Odor	Fruity	• the presence of ketones (breakdown of fat), which is a product of diabetes or starvation
	Foul	• the presence of bacteria
Transparency	Clear	• normal urine samples appear clear/transparent
	Cloudy	• old samples could appear cloudy if bacteria has had time to grow on it • fresh samples could appear cloudy if a urinary tract infection (UTI) is present (bacteria in the urethra) • fresh samples could appear cloudy if there are blood cells or pus
Sugar	Present	• patient ate a meal rich in carbohydrates prior to visit • a period of stress • diabetes
Protein	Present	• an abnormal condition called protein urea, that results from damage to kidneys

Chapter 3
Let's Diagnose Them, Lab 1: Urinalysis

Materials

- 8 test tubes (two labeled Patient #1, two labeled Patient #2, two labeled Patient #3, and two labeled Patient #4)
- 1 test tube rack
- 1 test tube clamp
- 10 ml graduated cylinder
- 1 cup containing 20 ml of Benedict's solution
- 1 cup containing 20 ml of Biuret reagent
- 2 droppers (one for Benedict, one for Biuret)
- 1 hot plate
- 1 250 ml beaker of water
- 1 set of goggles for EACH member of the team
- 6 ml urine samples from all four patients (retrieve from your teacher)

Procedure

Setup

1. Label two test tubes Patient #1, two test tubes Patient #2, two test tubes Patient #3, and two test tubes Patient #4.
2. Place 6 ml of each patient's urine sample into his or her designated test tubes.
3. Add 150 ml of water to your 250 ml beaker and preheat it on a hot plate *(needed later for the sugar test)*.

Physical Observations

1. Observe and describe the color, odor, and transparency of the four urine samples. Record physical descriptions in Table 3.2.
2. Get approval from teacher before advancing to Chemical Observations.

Chemical Observations

1. To test for sugar content, add approximately 6 ml of the TURQUOISE BLUE Benedict's solution to *one* test tube of *each* of the four patients' urine samples.

2. Place each patient's urine sample with the Benedict's solution into your hot water bath. Let sit for approximately five minutes.

3. If the TURQUOISE BLUE turns to ORANGE, as indicated by Data Table 3.3, sugar is present in the urine. Record whether or not sugar is present in Data Table 3.2.

4. To test for protein content, add approximately 2 ml of the BRIGHT BLUE biuret reagent into remaining test tubes for each of the urine samples. YOU DO NOT HEAT THESE TEST TUBES.

5. If the BRIGHT BLUE turns to a VIOLET PURPLE color, as indicated by Table 3.3, protein is present in the urine. Record whether or not protein is present in Table 3.2.

Chapter 3
Let's Diagnose Them, Lab 1: Urinalysis

TABLE 3.2. RESULTS OF EACH PATIENT'S URINALYSIS

	Patient #1	Patient #2	Patient #3	Patient #4
Urination Habits			N/A	N/A
Color • Dark yellow? • Pale yellow? • Red with blood?				
Odor • Fruity? • Foul? • Normal?				
Transparency • Cloudy? • Clear?				
Is sugar present? • Yes: Turned orange (with heat) • No: Did <u>not</u> turn orange (with heat)				
Is protein present? • Yes: Turned purple (with no heat) • No: Did <u>not</u> turn purple (with no heat)				

TABLE 3.3. MEANING OF NUTRIENT INDICATOR TEST RESULTS

Nutrient	If nutrient is not present, color will remain …	If nutrient is present, color will turn …
Glucose (sugar)	Turquoise Blue	(with heat) Orange
Protein	Bright Blue	(no heat) Violet Purple

Recall Questions

1. What is the role of the excretory system?

2. What is the function of the kidneys?

3. Why should a patient avoid eating a large meal before a urinalysis?

4. Why should a patient provide a fresh sample of urine opposed to a sample that has sat out for several days?

Critical Thinking Question

1. Dialysis is a medical device used to filter a patient's blood, when his or her kidneys can no longer function on their own, effectively. Dialysis is an expensive treatment and can cost as much as $500 per treatment. Medicare, the United States' medical insurance company for citizens ages 65 and older and for those with certain disabilities or kidney failure, covers most of the costs. However, under federal law, states are required to give emergency medical care to illegal immigrants, some of whom may require dialysis. As a result, taxpayers end up covering the cost. Considering your knowledge of the Hippocratic oath and the excretory system, determine whether you would support or oppose this federal law. Justify your position.

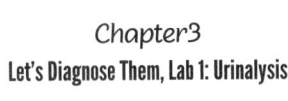

Chapter 3
Let's Diagnose Them, Lab 1: Urinalysis

Conclusion

1. Look back at your medical notes in Table 3.1, and your lab results in Table 3.2. What could these results indicate about your patients?

2. Return to your patients' medical charts (Tables 2.1, 2.2, 2.3, and 2.4, pp. 20–23 in Chapter 2) and complete the section labeled "Lab 1—Urinalysis" for each of the four patients. Check off evidence collected from each patient and consider whether or not your original hypothesis is still supported or refuted by evidence.

Chapter 4
Let's Diagnose Them, Lab 2
Digestive By-Products and Body Mass Index Analysis

Your patients have already spent one night at Vanguard Hospital and nurses on the nightshift had a very busy evening. The nurses just informed you that Patient #1 complained of excessive thirst, requesting water nearly every hour or so and Patient #2 was vomiting all morning and experiencing constipation. Additionally, Patient #3 experienced chronic episodes of diarrhea. The only patient with no major digestive issues was Patient #4; however, he did complain of abdominal pain. Nurses are concerned that Patients #2 and #3 are losing vital nutrients such as glucose, starch, protein, lipids, vitamins, minerals, and water. It will be essential to replace whatever nutrients are missing with intravenous fluids.

Today your team of doctors will run tests on the digestive by-products of Patient #2 (vomit) and Patient #3 (diarrhea). Since there was no by-product from Patient #1 or #4 you will not be performing any tests for them. However, you will be expected to formulate an even deeper hypothesis about the conditions of Patient #1 and #4 based on their symptoms.

Chapter 4
Let's Diagnose Them, Lab 2: Digestive By-Products and Body Mass Index Analysis

Lab Roles (Fill in Names of Team Members)

1. _____ is the task manager (reads procedure and ensures everyone is following proper protocol).

2. _____ is the materials manager (retrieves and returns materials; cleans materials and table).

3. _____ are the doctors (completes lab work, such as adding chemicals, heating chemicals, and so on; to be completed by more than one group member).

4. _____ is the recorder (ensures the group's data is properly recorded).

Materials

- 6 test tubes (3 labeled *Patient #2* and 3 labeled *Patient #3*)
- 1 test tube rack
- 1 test tube clamp
- 10 ml graduated cylinder
- 1 cup containing 20 ml of Benedict's solution
- 1 cup containing 20 ml of biuret reagent
- 1 cup containing 20 ml of Lugol's solution (iodine)
- 3 droppers (one for Benedict, one for biuret, one for Lugol's)
- 1 hot plate
- 250 ml beaker of water
- 1 set of goggles (for EACH team member)
- 6 ml samples of digestive by-products from Patients #2 and #3 (retrieve from your teacher)

Chapter 4
Let's Diagnose Them, Lab 2: Digestive By-Products and Body Mass Index Analysis

Procedure

Setup

1. Label three test tubes *Patient #2* and three test tubes *Patient #3*.
2. Place 6 ml of each patient's by-product into his or her designated test tubes.
3. Add 150 ml of water to your 250 ml beaker and preheat it on a hot plate *(needed later for the sugar test)*.

Physical Observations

1. Observe and describe the color and texture of the two patients' digestive by-products.
2. Record physical observations in Table 4.1.
3. Get approval from teacher before advancing to Chemical Observations.

Chemical Observations

1. To test for sugar content, add 6 ml of the TURQUOISE BLUE Benedict's solution to one test tube for each of the patient's digestive by-products.
2. Place each patient's digestive by-product with the Benedict's solution into your hot water bath. Let sit for approximately five minutes.
3. If the TURQUOISE BLUE turns to ORANGE, as indicated by Table 4.2, sugar is present in the digestive by-product. Record whether or not sugar is present in Table 4.1.
4. To test for protein content, add approximately 2 ml of the BRIGHT BLUE biuret reagent remaining in the test tubes for each of the patients' digestive by-product samples. YOU DO <u>NOT</u> HEAT THESE TEST TUBES.
5. If the BRIGHT BLUE turns to a VIOLET PURPLE color, as indicated by Table 4.2, protein is present in the digestive by-product. Record whether or not protein is present in Table 4.1.
6. To test for starch content, add approximately five drops of the AMBER BROWN Lugol's solution into the leftover test tubes for each of the patients' digestive by-product. Mix solution. YOU DO <u>NOT</u> HEAT THESE TEST TUBES.
7. If the AMBER BROWN turns to a PURPLE/BLACK color, as indicated by Table 4.2, starch is present in the digestive by-product. Record whether or not starch is present in Table 4.1.

TABLE 4.1. NUTRITION CONTENT FOUND IN PATIENT #2 AND #3'S DIGESTIVE BY-PRODUCTS

	Patient #1	Patient #2	Patient #3	Patient #4
Physical observations of by-products	Given: • thirsty all night • no vomit or diarrhea			Given: • abdominal pain • no digestive problems
Glucose • Present or not present?	N/A			N/A
Protein • Present or not Present?	N/A			N/A
Starch • Present or not Present?	N/A			N/A

TABLE 4.2. MEANING OF NUTRIENT INDICATOR RESULTS

Nutrient	If nutrient is not present, color will remain …	If nutrient is present, color will turn …
Glucose (sugar)	Turquoise Blue	(with heat) Orange
Protein	Bright Blue	(no heat) Violet Purple
Starch	Amber/Brown	(no heat) Purple/Black

Chapter 4
Let's Diagnose Them, Lab 2: Digestive By-Products and Body Mass Index Analysis

FIGURE 4.1. BODY MASS INDEX (BMI) CHART FOR ADULTS

Weight in Pounds (lbs)

Height (ft.)	100	105	110	115	120	125	130	135	140	145	150	155	160	165	170	175	180	185	190	195	200
5'	19	20	21	22	23	24	25	26	27	28	29	30	31	32	33	34	35	36	37	38	39
5'1"	18	19	20	21	22	23	24	25	26	27	28	29	30	31	32	33	34	35	36	36	37
5'2"	18	19	20	21	22	22	23	24	25	26	27	28	29	30	31	32	33	33	34	35	36
5'3"	17	18	19	20	21	22	23	24	24	25	26	27	28	29	30	31	32	32	33	34	35
5'4"	17	18	18	19	20	21	22	23	24	24	25	26	27	28	29	30	31	31	32	33	34
5'5"	16	17	18	19	20	21	21	22	23	24	25	25	26	27	28	29	30	30	31	32	33
5'6"	16	17	17	18	19	20	21	21	22	23	24	25	25	26	27	28	29	30	31	32	33
5'7"	15	16	17	18	18	19	20	21	22	22	23	24	25	25	26	27	28	29	29	30	31
5'8"	15	16	17	18	18	19	19	20	21	22	23	24	25	25	26	27	28	28	29	30	31
5'9"	14	15	16	17	17	18	19	20	20	21	22	22	23	24	25	25	26	27	28	28	29
5'10"	14	15	15	16	17	18	18	19	20	20	21	22	23	24	24	25	25	26	27	28	28
5'11"	14	14	15	16	16	17	18	18	19	20	21	21	22	23	23	24	25	25	26	27	27
6'	13	14	14	15	16	16	17	18	18	19	20	21	21	22	23	24	25	25	26	27	27
6'1"	13	13	14	15	15	16	17	17	18	19	19	20	21	21	22	23	23	24	25	25	26
6'2"	12	13	14	14	15	16	16	17	18	18	19	19	20	21	21	22	23	23	24	25	25

TABLE 4.3. BMI RANGES

Underweight	Healthy	Overweight	Obese
<18.9	19–24.9	25–29.9	>30

Chapter 4
Let's Diagnose Them, Lab 2: Digestive By-Products and Body Mass Index Analysis

Nurses also mentioned they are concerned about the weight loss and gain the patients are experiencing. Some appear to have lost a significant amount of weight over a short period of time and some appear to be severely overweight or gaining weight at a rapid rate.

Body Mass Index (BMI)

1. Locate your patient's height (in feet and inches) and weight (in pounds) provided on each medical record (pages 12–19 in Chapter 2).
2. Using Figure 4.1, determine the BMI number for each of your patients.
3. Similarly, determine if your patients are underweight, overweight, obese, or healthy for their height using Table 4.3.
4. Record your findings in Table 4.4.

TABLE 4.4. THE BMI VALUE FOR EACH PATIENT

	Patient #1	Patient #2	Patient #3	Patient #4
Height				
Weight				
BMI #				
Are they ... Healthy? Underweight? Overweight? Obese?				

Recall Questions

1. What is the role of the digestive system?

2. What is the difference between simple carbohydrates and complex carbohydrates?

Chapter 4
Let's Diagnose Them, Lab 2: Digestive By-Products and Body Mass Index Analysis

3. What is the glycemic index (GI)?

4. What foods are considered high on the glycemic index? What foods are considered low?

5. Why is a high BMI value considered worrisome for doctors?

6. Explain why BMI charts, although used by doctors, may not accurately depict one's true state of health.

Critical Thinking Question

1. During summer 2012, in an attempt to curb childhood obesity, the U.S. Department of Food and Agriculture mandated schools to offer nutritious meals at breakfast and lunch. Some of the guidelines drafted include:

    - offering vegetables, fruit, whole-grains, meat, meat alternatives, and fat-free unflavored milk daily and at each meal time
    - reducing the sodium content of meals over a 10-year period
    - preparing meals that contain 0 grams of trans fats, and
    - designing meals that target the specific caloric needs of varying age groups

Suppose you were selected to advise the Department of Food and Agriculture; using your knowledge of nutrition and its effects on a person's health, construct a five-day meal plan, for *either* breakfast or lunch, that would be considered acceptable under these new regulations. Be sure to include the nutrition information in the meal plan (amounts of sodium, trans fats, carbohydrates, and fiber; total calories; and so on).

Conclusion

1. Look back at your lab results in Table 4.1 and Table 4.4. What could these results indicate about your patients?

2. Return to your patients' medical charts (Tables 2.1, 2.2, 2.3, and 2.4, pp. 20–23 in Chapter 2) and complete the section labeled "Lab 2—Digestive By-Products and BMI Analysis" for each of the four patients. Check off evidence collected from each patient and consider whether or not your original hypothesis is still supported or refuted by evidence.

Chapter 5
Let's Diagnose Them, Lab 3
Blood Smears

Some doctors on your team are beginning to think that some of your patients' symptoms may be caused by either a pathogen or a genetic disorder. A pathogen causes harm or disease in another living organism. Examples include viruses, bacteria, and fungi. Genetic disorders are diseases inherited from one's parents.

Today your team of doctors will analyze the red blood cells (RBCs) of patients under a microscope. Nurses have also provided you with your patients' red blood cell and white blood cell counts. Use your medical school notes (Table 5.1) as a reference for diagnosing your patients.

TABLE 5.1. MEDICAL SCHOOL NOTES REGARDING RED AND WHITE BLOOD CELLS

	Function	Healthy if ...	Unhealthy if ...
Red blood cells (RBC)	Uses the protein, hemoglobin, to carry oxygen around the body	Shaped like a donut Female RBC count = 4.2–5.4 million/μL/cu mm Male RBC count = 4.7–6.1 million/μL/cu mm	Shaped like a sickle, indicating a genetic disorder called sickle cell anemia. If *lower* than normal, could indicate anemia, such as sickle cell anemia. However, anemia is also common during the first six months of pregnancy If *higher* than normal, could indicate polycythaemia, a disorder of the bone marrow.
White blood cells (WBC)	Help fight infections by (A) Phagocytosis of foreign agents (B) Producing antibodies against foreign agents	WBC count = 4,300–10,800 cells/μL/cu mm	If *lower* than normal, could indicate viral infections like HIV, low immunity and bone marrow failure If *higher* than normal, could indicate infection, systemic illness, inflammation, allergy, leukemia, and tissue injury caused by burns, or pregnancy.

Chapter 5
Let's Diagnose Them, Lab 3: Blood Smears

Lab Roles (Fill in Names of Team Members)

1. _____ is the task manager (reads procedure and ensures everyone is following proper protocol).

2. _____ is the materials manager (retrieves and returns materials; cleans materials and table).

3. _____ are the doctors (completes lab work, such as adding chemicals, heating chemicals, and so on; to be completed by more than one group member).

4. _____ is the recorder (ensures the group's data is properly recorded).

Materials

- 1–2 compound light microscopes
- Blood smears from your four patients, provided by your teacher

Procedure

1. Start with the microscope stage as far away from the lens as possible.
2. Place Patient #1's blood smear on the stage and secure it with the stage clips.
3. Place the objective lens to low power (4×).
4. Using the coarse adjustment (big knob), begin to focus the slide.
5. Once focused, change the objective lens to medium power (10×).
6. Using the coarse adjustment (big knob), begin to focus the slide.
7. Once focused, change the objective lens to high power (40×).
8. Using the fine adjustment (small knob), begin to focus the slide.
9. Sketch your observation of red blood cells at the power most easily observable in Table 5.2.
10. Repeat steps 1–9 for patients #2, #3, and #4.

Chapter 5
Let's Diagnose Them, Lab 3: Blood Smears

TABLE 5.2. THE BLOOD SMEAR RESULTS FOR EACH PATIENT

	Patient #1	Patient #2	Patient #3	Patient #4
Sketch a detailed picture of what you observe here:				
RBC shape: Normal or sickle?				
# of RBCs million/µL/cu mm Normal, high, or low?	4.4	3.0	5.1	3.2
# of WBCs /µL/cu mm Normal, High, or low?	7,004	11,300	2,029	9,001

Recall Questions

1. What is the role of the circulatory system?

2. How do arteries differ from veins?

3. What problems could arise in the circulatory system from poor nutrition and lack of exercise?

4. What is the role of a red blood cell?

5. How does sickle cell anemia differ from sickle cell trait?

6. What happens to the hemoglobin protein on a red blood cell if someone has sickle cell anemia?

Critical Thinking Question

1. In early 2012, coaches instructed a Pittsburgh professional football player, with the sickle cell *trait*, to sit out a game in the high-altitude city, of Denver, Colorado. Doctors claimed that the trait, in combination with extreme physical activity and a high altitude, was the primary reason he needed to have his spleen and gallbladder removed after a previous game in the city. However, it has been estimated that at least 90 other NFL players carry the sickle cell trait, and of those who have played in Denver, they have never experienced such issues before. In fact, a study performed by Howard University in 2000, showed no complications in athletes carrying sickle cell trait during the Mexico City Olympics, another high-altitude location. Suppose you were a coach of a high school, college, or professional sports team. Knowing what you know about sickle cell anemia and the trait, how would you handle a situation similar to this one, in which one of your players has sickle cell trait or the disease? Justify your position.

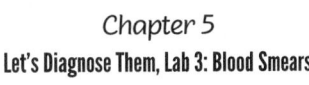

Chapter 5
Let's Diagnose Them, Lab 3: Blood Smears

Conclusion

1. Look back at your medical notes in Table 5.1, and your lab results in Table 5.2. What could these results indicate about your patients?

2. Return to your patients' medical charts (Tables 2.1, 2.2, 2.3, and 2.4, pp. 20–23 in Chapter 2) and complete the section labeled "Lab 3—Blood Smears" for each of the four patients. Check off evidence collected from each patient and consider whether or not your original hypothesis is still supported or refuted by evidence.

Chapter 6
Let's Diagnose Them, Lab 4
HIV Test

Based on blood samples recently analyzed, one of your patients, Patient #3, had a very low white blood cell (WBC) count. This is very alarming to you and your team of doctors since it might indicate the patient is suffering from a viral infection. Since all of your patients have requested an STD test on their medical records, you will specifically check for antibodies produced against the Human Immunodeficiency Virus (HIV).

Before you test your patients, it is essential that you discuss the importance of safe sex. As you have already noticed, all patients indicated on their medical records that they are currently sexually active, yet none of them reported the regular use of condoms as a method of protection. In order to emphasize the importance of safe sex, you and the other doctors in the hospital will demonstrate a simulation of how fast an STD can travel within a population of multiple sex partners who are engaging in unsafe sex. Upon completing this "sex-talk"/demonstration, you will then test your patients.

Chapter 6
Let's Diagnose Them, Lab 4: HIV Test

Lab Roles (Fill in Names of Team Members)

1. _____ is the task manager (reads procedure and ensures everyone is following proper protocol).

2. _____ is the materials manager (retrieves and returns materials; cleans materials and table).

3. _____ are the doctors (completes lab work, such as adding chemicals, heating chemicals, and so on; to be completed by more than one group member).

4. _____ is the recorder (ensures the group's data is properly recorded).

Materials

Part I

- 4 cups per doctor (one filled with NaOH ("virus") and three with H_2O ("no virus") designated by the teacher
- Phenolphthalein (the "HIV indicator")
- 1 dropper
- Distilled water
- 1 set of goggles

Part II

- 4 test tubes (labeled Patient #1, Patient #2, Patient #3, and Patient #4, respectively)
- 1 cup with phenolphthalein (the "HIV indicator")
- 1 dropper
- Bodily fluid samples from all four patients

Chapter 6
Let's Diagnose Them, Lab 4: HIV Test

Procedure

Part I: Done as a Class

ROUND 1: Multiple Sexual Partners

1. Obtain two cups labeled "Control 1" and "Round 1" from your teacher. The fluids inside the two cups are the same, and represent vaginal fluid and/or semen. Everyone in the class will have H_2O ("no virus") in their two cups except for *one* person who will have NaOH ("virus").

2. When your teacher says "Go!" find *five* different people to have "unprotected sex" with, represented by the exchange of fluids in your "Round 1" cup. Make sure your Control 1 cup is off to the side and remains uncontaminated.

3. Return to your seat after changing fluids with five different people.

4. Upon completion, the head doctor (your teacher) will give each of you an HIV test. The head doctor will place 1 drop of phenolphthalein (the "HIV indicator") into your Round 1 cups.

5. If fluid in your Round 1 cup remains clear, you did *not* contract HIV. If the fluid turns pink, you did contract HIV. Record your observations in Table 6.1.

6. Who initiated the infection? The head doctor will place 1 drop of phenolphthalein (the "HIV indicator") into your Control cup. Record your observations in Table 6.1.

7. Dispose of your cups.

Chapter 6
Let's Diagnose Them, Lab 4: HIV Test

TABLE 6.1. RESULTS FROM MULTIPLE SEX PARTNER DEMONSTRATION

Color of the fluid in my Round 1 cup after HIV test:	
Total number of "doctors" with HIV after exchanging fluids with five different partners:	
The person who initially transmitted the HIV infection to everyone in the room was:	

ROUND 2: Limited Sexual Partners

1. Obtain two new cups (labeled "Control 2" and "Round 2") from your teacher. Again, the fluids inside the two cups are the same, and represent vaginal fluid and/or semen. Everyone in the class will have H_2O ("no virus") in their two cups except for *one* person who will have NaOH ("virus").

2. When your teacher says "Go!" find *two* different people to have "unprotected sex" with, represented by the exchange of fluids in your Round 2 cups. Make sure your Control 2 cup is off to the side and remains uncontaminated.

3. Return to your seat after exchanging fluids with two different people.

4. Upon completion, the head doctor (your teacher), will give each doctor an HIV test. The head doctor will place 1 drop of phenolphthalein (the "HIV indicator") into your Round 2 cup.

5. If the fluid in your Round 2 cup remains clear, you did *not* contract HIV. If the fluid turns pink, you did contract HIV. Record your observations in Table 6.2.

6. Who initiated the infection? The head doctor will place 1 drop of phenolphthalein (the "HIV indicator") into your Control cup. Record your observations in Table 6.2.

7. Dispose of your cups.

TABLE 6.2. RESULTS FROM LIMITED SEX PARTNER DEMONSTRATION

Color of the fluid in my Round 2 cup after test:	
Total number of "doctors" with HIV after exchanging fluids with two partners?	
The person who initially transmitted the HIV infection to everyone in the room was:	

Chapter 6
Let's Diagnose Them, Lab 4: HIV Test

Part II: Done as a Team of Doctors

1. Label four test tubes *Patient #1, Patient #2, Patient #3,* and *Patient #4,* respectively.

2. Obtain bodily fluids from each patient and place the samples in their designated test tubes.

3. Perform the HIV test on each of them by adding 1 drop of phenolphthalein (the "HIV indicator") to their semen or vaginal fluid sample.

4. Record the color in Table 6.3 and determine whether or not they have HIV.

TABLE 6.3. RESULTS FROM EACH PATIENT'S HIV TEST

	Patient #1	Patient #2	Patient #3	Patient #4
Color • Clear or Pink?				
HIV status • HIV + or –?				

Chapter 6
Let's Diagnose Them, Lab 4: HIV Test

Recall Questions

1. What is the role of the immune system?

2. What is the role of a white blood cell?

3. At what point does HIV become AIDS?

4. Why does someone with AIDS become more susceptible to other infections?

5. What are some ways a person can prevent the spread of HIV to others?

6. List one difference between a bacterial infection and a viral infection.

7. Approximately how long after the initial infection will HIV antibodies show up in an HIV test?

8. What is a vaccine and how does it work?

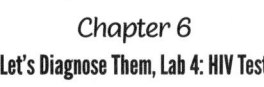

Critical Thinking Question

1. At the moment, HIV tests are not included in a routine doctor check-up and must be requested by the patient. Should HIV tests become routine and mandated for all sexually active individuals and/or individuals 18 and older? Explain your position.

Conclusion

1. Look back at your lab results in Table 6.1 and Table 6.2. What can you conclude about the relationship between the number of sexual partners (without protection) and the risk of receiving a sexually transmitted disease?

2. Look back at your lab results in Table 6.3. What do these results indicate about your patients?

3. Return to your patients' medical charts (Tables 2.1, 2.2, 2.3, and 2.4. pp. 20–23 in Chapter 2) and complete the section labeled "Lab 4—HIV Test" for each of the four patients. Check off evidence collected from each patient and consider whether or not your original hypothesis is still supported or refuted by evidence.

Chapter 7
Let's Diagnose Them, Lab 5
Lung Capacity

All four patients have been at Vanguard Hospital for the last two days. Nurses report that all of them are experiencing difficulty breathing. In medical school, you learned the average pair of human lungs can hold about 5 liters or 5,000 cubic centimeters (cc) of air, but only a small amount of this capacity is used during normal breathing (roughly 1,000 cc). Today you will determine the lung capacity of each of your patients. Lung capacity is the maximum amount of air the lungs can hold. Normally, doctors use a spirometer to determine a patient's lung capacity. A spirometer requires the patient to exhale deeply in order to determine if diseases such as asthma, pneumonia, and bronchitis are compromising the patient's respiratory system.

To refresh your memory of lung capacity, let's take a peek at your class notes (Table 7.1) from medical school:

TABLE 7.1. FACTORS THAT MAY INCREASE OR DECREASE LUNG CAPACITY

People With Larger Volumes and Unrestricted Breathing	People With Smaller Volumes and Compromised Breathing
Males	Females
Taller people	Shorter people
Nonsmokers	Smokers
Athletes	Non-athletes
People living at high altitudes	People living at low altitudes
Nonpregnant women	Pregnant women
Healthy weight	Obesity
Normal red blood cells	Sickle cell anemia
Healthy respiratory tracts	Restricted respiratory tracts

Chapter 7
Let's Diagnose Them, Lab 5: Lung Capacity

As can be seen from the medical school notes, a variety of factors may impact one's lung capacity. Women who are pregnant, for instance, often experience smaller lung volumes since the growing baby pushes up on the diaphragm from the uterus. Similarly, individuals with sickle cell anemia struggle with their breathing since the hemoglobin protein on their RBCs are mutated and only carry half the number of oxygen molecules as a normal RBC. Diabetics also exhibit compromised lung volumes due to high blood sugar levels stiffening the lung tissue and fatty tissue in the abdominal area. And as mentioned previously, those with HIV tend to suffer from opportunistic infections such as pneumonia, which causes the lungs to fill up with mucous.

Today you will compare each of your patients' lung capacities (from balloon blowing) to the expected lung capacity of someone with the same height, age, and gender. Under normal circumstances, doctors might ask their patients to exhale into a spirometer.

Chapter 7
Let's Diagnose Them, Lab 5: Lung Capacity

Lab Roles (Fill in Names of Team Members)

1. _____ is the task manager (reads procedure and ensures everyone is following proper protocol).

2. _____ is the materials manager (retrieves and returns materials; cleans materials and table).

3. _____ are the doctors (completes lab work, such as adding chemicals, heating chemicals, and so on; to be completed by more than one group member).

4. _____ is the recorder (ensures the group's data is properly recorded).

Materials

- 4 inflated balloons of different diameters
- 1 wind-up measuring tape
- 1–2 scientific calculators

Procedure

Part I: Patient's Lung Capacity

1. Find the circumference for Patient #1's balloon by wrapping your roll-up ruler around the widest portion of the balloon. Measure the length in centimeters. Record this value in Table 7.2.

2. Repeat step 1 for Patient #2, Patient #3, and Patient #4.

3. Using the formula for circumference, find the radius (r) of the balloon. Plug in the value for C and solve for r. Remember: The value of π is 3.14. Record the value of r in Table 7.2.

> **Circumference Equation**
> $C = 2\pi r$
> $r = C/2\pi$

4. Using the radius you just solved for, determine the diameter of the balloon. Plug in the value for r and solve for d. Record the value of d in Table 7.2.

> **Diameter Equation**
> $d = 2r$

5. On the x-axis of Figure 7.1, locate the diameter of the balloon in centimeters and follow the number up until it meets the curved line. Then move across, in a straight line, to the vertical y-axis. Approximate the lung volume for your patient. Record the lung volume in Table 7.2.

6. Repeat steps 3–5 for patients #2, #3 and #4.

TABLE 7.2. CALCULATIONS FOR DETERMINING YOUR PATIENTS' LUNG CAPACITIES

	Patient #1	Patient #2	Patient #3	Patient #4
Measure the circumference in centimeters (cm)				
Calculate the radius in centimeters (cm) $r = C/2\pi$				
Calculate the diameter in centimeters (cm) $d = 2r$				
Determine your patient's lung capacity (cc) using the graph				

FIGURE 7.1. DETERMINING LUNG CAPACITY BY A BALLOON'S DIAMETER

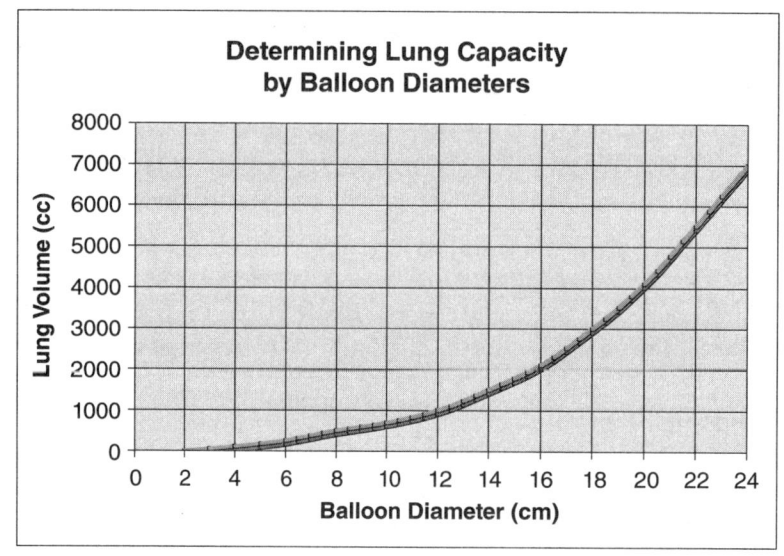

Chapter 7
Let's Diagnose Them, Lab 5: Lung Capacity

Part II: Acceptable Lung Capacity for Height, Weight, and Gender

Research has shown the capacity of a person's lungs *should* be proportional to the surface area of his or her body. To find the surface area of your patients, you will need to know the height, weight, and gender of each, which are listed in their medical records. There are a couple of different ways to calculate mathematically a person's body surface area and estimate their *acceptable* lung capacity, mathematically.

1. To determine the acceptable lung capacity of your patients, enter their heights, weights, and gender values into the Body Surface Area equation below. Record this value in Table 7.3. *Note:* Height must be in centimeters and weight must be in kilograms. Your head doctor (the teacher) has completed these conversions for you already. Use the **bold** values for your calculations.

$$\text{Body Surface Area} = \sqrt{(\text{height in cm} \times \text{weight in kg})/3600}$$

If *Female* = Body Surface Area Value × 2000

If *Male* = Body Surface Area Value × 2500

TABLE 7.3. DETERMINING PATIENTS' ACCEPTABLE LUNG CAPACITIES

	Sex	Height (ft → cm)	Weight (lbs → kg)	Acceptable Lung Capacity (cc) (Show math!)
Patient #1	F	5'6" → **167.7 cm**	200 lbs → **90.7 kg**	
Patient #2	F	5'2" → **157.5 cm**	120 lbs → **54.5 kg**	
Patient #3	M	6'2" → **188.0 cm**	140 lbs → **63.5 kg**	
Patient #4	M	5'8" → **172.7 cm**	155 lbs → **70.3 kg**	

2. Compare the results you obtained in Tables 7.2 and 7.3 and complete Table 7.4.

TABLE 7.4. COMPARISON OF EACH PATIENT'S ACTUAL LUNG CAPACITY TO HIS OR HER ACCEPTABLE LUNG CAPACITY

	Patient #1	Patient #2	Patient #3	Patient #4
How does your patient's lung capacity compare to the acceptable lung capacity of someone with the same height, weight, and gender (higher, lower, or similar)?				

Recall Questions

1. What is the role of the respiratory system?

2. When a person exhales, what happens to the diaphragm? What happens during an inhale?

3. What happens in the alveoli?

4. Why would a pregnant person have a lower lung capacity?

5. Why might a person with sickle cell anemia have difficulty breathing?

Chapter 7
Let's Diagnose Them, Lab 5: Lung Capacity

6. Why would someone who has diabetes have difficulty breathing?

7. Why would someone who has HIV have difficulty breathing?

Critical Thinking Question

1. Asthma is a respiratory disease characterized by restricted airflow, resulting in difficulty breathing and a variety of other chest-related symptoms. Asthma attacks can be brought on by allergies, but are often brought on by poor air quality. In 2012 the U.S. Centers for Disease Control and Prevention stated that asthma cases rose from 7.3% in 2001 to 8.4% in 2010 and were higher among children, than adults, and among multiple-race, black, and American Indian or Alaska Native persons than white persons. In particular, low-income urban areas tended to have higher asthmatic cases. Considering your knowledge of respiratory health, is it the government's responsibility to make asthma prevention a priority or should business and factories be held accountable for improving outdoor air quality? Create a proposal that would satisfy the needs of urban residents, government officials, and big businesses.

Chapter 7
Let's Diagnose Them, Lab 5: Lung Capacity

Conclusion

1. Look back at your lab results in Table 6.3. What could these results indicate about your patients?

2. Return to your patients' medical charts (Tables 2.1, 2.2, 2.3, and 2.4. pp. 20–23 in Chapter 2) and complete the section labeled "Lab 5—Lung Capacity" for each of the four patients. Check off evidence collected from each patient and consider whether or not your original hypothesis is still supported or refuted by evidence.

Chapter 8
Let's Diagnose Them, Lab 6
Hormone Test

Your patients have been in this hospital for three days. By now, you have most likely verified a diagnosis for your male patients, Patients #3 and #4. You also might suspect that one of your female patients is pregnant and that the other is diabetic.

Every doctor knows the endocrine system produces hormones that help regulate the body's internal organs. Hormones are chemical messengers sent throughout the bloodstream. When too little or too much of a hormone is produced, it is often an indication that something is wrong.

In medical school (Table 8.1), you learned the following about two hormones produced in the body:

TABLE 8.1. MEDICAL SCHOOL NOTES REGARDING TWO HORMONES: HCG AND INSULIN

	Increased levels could indicate …	Decreased levels could indicate …
Human Chorionic Gonadotrophin (hCG)	• Pregnancy only	• Not pregnant
Insulin	• Drugs such as corticosteroids, levodopa, and oral contraceptives • Fructose or galactose intolerance • Excessive exercising	• Diabetes • Pancreatic diseases such as chronic pancreatitis and pancreatic cancer

To determine if someone is diabetic, a doctor must provide patients with a glucose tolerance test. This test records how quickly sugar is cleared from

Chapter 8
Let's Diagnose Them, Lab 6: Hormone Test

the blood stream. The test is most frequently used to determine if a person is diabetic. The patient in question is required to fast 8 to 14 hours before they take the test. Only water is allowed. The patient is then given a glucose solution to drink. Blood is drawn at different intervals, and glucose levels are measured each hour. The glucose levels following the 2-hour mark are the most critical in determining if a person is diabetic. Glucose levels above 200 mg/dl show that insulin levels are low, suggesting diabetes.

Now that you have reviewed your medical school knowledge, let's diagnose these patients!

Chapter 8
Let's Diagnose Them, Lab 6: Hormone Test

Lab Roles (Fill in Names of Team Members)

1. _____ is the task manager (reads procedure and ensures everyone is following proper protocol).

2. _____ is the materials manager (retrieves and returns materials; cleans materials and table).

3. _____ are the doctors (completes lab work, such as adding chemicals, heating chemicals, and so on; to be completed by more than one group member).

4. _____ is the recorder (ensures the group's data is properly recorded).

Materials

- 2 test tubes
- 1 test tube rack
- blue litmus paper ("pregnancy" test)
- 3 ml urine samples from Patients #1 and #2 (retrieve from your teacher)

Procedure

Part I: Pregnancy Test

1. Label two test tubes *Patient #1* and *Patient #2*.

2. Obtain a 3 ml urine sample from both patients. Be sure to place them into their designated test tubes.

3. Dip the BLUE litmus paper ("pregnancy test") into the urine sample of Patient #1. If the pregnancy test does not change colors, as indicated in Table 8.2, the patient has normal levels of hCG in their urine and is not pregnant. If the pregnancy test changes to a PINK color, the patient has high levels of hCG in their urine and is therefore pregnant.

4. Record data for Patient #1 in Table 8.3.

5. Repeat steps 3–4 for Patient #2.

TABLE 8.2. MEANING OF HCG INDICATOR RESULTS

If color does not change, it indicates …	If color changes, it indicates …
No pregnancy	Pregnancy

Chapter 8
Let's Diagnose Them, Lab 6: Hormone Test

TABLE 8.3. RESULTS OF PATIENT #1'S AND #2'S PREGNANCY TESTS

	Patient #1	Patient #2	Patient #3	Patient #4
Pregnancy test • hCG Present?			male (no pregnancy test done)	male (no pregnancy test done)

Part II: Glucose Tolerance Test

1. Interpret the Glucose Tolerance Test results (Figure 8.1) for all four patients.
2. Use your medical school notes (Table 8.4) to determine which patients are normal or diabetic.
3. Record results in Table 8.5.

TABLE 8.4. MEDICAL SCHOOL NOTES REGARDING NORMAL AND DIABETIC TEST RESULTS FOR A GLUCOSE TOLERANCE TEST

Glucose levels	Normal		Diabetic	
Venous Blood Plasma	Fasting	2 hrs	Fasting	2 hrs
(mg/dl)	<110	<140	>126	>200

FIGURE 8.1. RESULTS FOR EACH PATIENT'S GLUCOSE TOLERANCE TEST

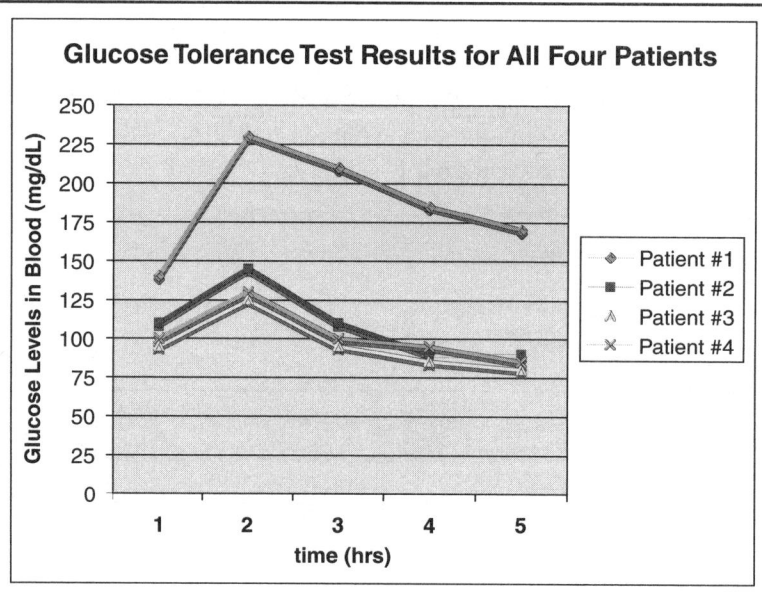

TABLE 8.5. RESULTS FOR EACH PATIENT'S GLUCOSE TOLERANCE TEST

	Patient #1	Patient #2	Patient #3	Patient #4
Glucose level after two hours: Above 200 mg/dl? Between 145–200 mg/dl? Below 145 mg/dl?				
Insulin level must be: Low? Borderline? Normal?				

Recall Questions

1. What is the role of the reproductive system?

2. What is the role of the endocrine system?

3. Why is hCG only found in pregnant women?

4. What is the role of insulin?

5. What organ produces insulin?

6. Explain how insulin and glycogen are examples of a feedback loop. In other words, how do they help maintain homeostasis in one's body?

7. What is the difference between type 1 and type 2 diabetes?

Critical Thinking Question

1. According to Guttmacher Institute's "State Policies in Brief" (2012), "37 states require parental involvement in a minor's decision to have an abortion" (p. 1). The term *parental involvement* however, ranges in definition. For some states, it refers to parental consent or others' notification, and for some, a notarized document. Additionally, the need for parental involvement varies case by case (i.e., medical emergency, abuse/assault/incest/neglect, and so on). Suppose you were the doctor of a 16-year-old pregnant patient in a state that did *not* mandate parental involvement, what medical advice might you provide to your patient to help her make an informed decision moving forward?

Conclusion

1. Look back at your lab results in Table 8.3 and Table 8.5. What could these results indicate about your patients?

2. Return to your patients' medical charts (Tables 2.1, 2.2, 2.3, and 2.4. pp. 20–23 in Chapter 2) and complete the section labeled "Lab 6—Hormone test" for each of the four patients. Check off evidence collected from each patient and consider whether or not your original hypothesis is still supported or refuted by evidence.

Chapter 9
Emergency! Lab 7
Performing Surgery

Fortunately, you have been taking great care of your patients. However, a fifth patient (represented by the rat) has just been rushed into the emergency room at Vanguard Hospital. Doctors have been working on this patient, tirelessly, but were unsuccessful reviving him/her. The patient has indicated that he/she is an organ donor. Doctors in the operating room have called upon you to perform the donation surgery. Given your knowledge of anatomy and physiology from medical school and your work at Vanguard Hospital, it will be your job to identify and remove various organs from the patient's body.

Chapter 9
Emergency! Lab 7: Performing Surgery

Lab Roles (Fill in Names of Team Members)

1. _____ is the task manager (reads procedure and ensures everyone is following proper protocol).

2. _____ is the materials manager (retrieves and returns materials; cleans materials and table).

3. _____ are the doctors (completes lab work, such as adding chemicals, heating chemicals, and so on; to be completed by more than one group member).

4. _____ is the recorder (ensures the group's data is properly recorded).

Materials (Per Groups of Four)

- 1 rat
- 1 dissection tray
- 1 scalpel/dissection scissor
- several dissection pins
- 1 dropper
- gloves for each person handling the rat
- goggles for each team member
- 1 biology textbook for reference

Procedure

1. Make sure all doctors (students) are wearing proper safety gear (gloves, goggles)
2. Lay the rat on its back.
3. Pierce the rat's abdomen with the scalpel/dissection scissor. Cut vertically (Figure 9.1) from the top of the abdomen area to the pelvic region.
4. At the top of the abdomen and at the bottom of the abdomen, cut horizontally. There should now be two flaps that open like a book. Pin these flaps back with the dissection pins.

FIGURE 9.1. PREPARING FOR RAT DISSECTION

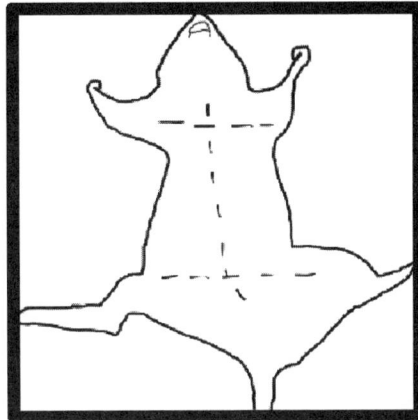

5. Right now you are looking inside the rat's abdomen area.

    • **Question 1:** What organ takes up most of the space in the rat's abdomen?

    • **Question 2:** Identify the stomach. Sketch it below.

    • **Question 3:** What tube is leading toward the stomach?

    • **Question 4:** What tube is connected to the bottom of the stomach?

Chapter 9
Emergency! Lab 7: Performing Surgery

6. Cut out the stomach. Open it up.

    • **Question 5:** What does the stomach look like inside? Describe it.

7. Gently pull out the small intestine.

    • **Question 6:** How long is the small intestine? Measure its length in centimeters.

    • **Question 7:** What is the job of the small intestine?

8. Gently pull out the large intestine.

    • **Question 8:** How long is the large intestine? Measure its length in centimeter.

    • **Question 9:** What is the job of the large intestine?

9. Open the large intestine.

    • **Question 10:** What is inside of it?

10. At this point, you have already removed a large portion of the organs in the abdominal cavity of the rat. Located toward the back of the rat are two bean like structures, called the kidneys. See if you can find them.

    • **Question 11:** What is the function of the kidneys?

11. Open up one of the kidneys.

    • **Question 12:** What does the kidney look like inside?

12. The kidneys are attached to two tubes called the ureter. The ureter connects to the bladder.

    • **Question 13:** What is the role of the bladder?

13. Since you are now in the pelvic region of the rat, see if you can determine the gender of your rat.

    • **Question 14:** What gender is your rat? How do you know? What did you find—or not find—that may support your answer?

Chapter 9
Emergency! Lab 7: Performing Surgery

14. Now change gears. Let's concentrate on the upper region of the rat that hasn't been exposed. At the very top there should be a membrane-like muscle separating the upper abdomen from the chest region of the rat.

    • **Question 15:** What is the name of this muscle that is located below the ribs that spans the width of the rat?

15. Make a vertical cut through the chest of the rat. You will probably need to use a little bit more force to break through the ribs. Once cut, open up the chest cavity. Here you should see the lungs and the heart.

    • **Question 16**: Sketch the structure of the lungs and heart below.

    • **Question 17:** Explain how the heart and lungs work together to help your body function properly.

16. If you glance around the rat, you probably notice hot pink blood vessels and blue blood vessels. The rats have been injected with a serum to help identify the arteries and veins.

    • **Question 18:** What is the job of the arteries?

    • **Question 19**: What is the job of the veins?

- **Question 20:** Use your knowledge of the circulatory system to explain why the hot pink and blue blood vessels are misleading.

17. Try to identify the pipe leading from the mouth to the lungs.

    - **Question 21:** What is the name of this pipe?

18. Job well done! Make sure you clean up and wipe down your area. Nurses were able to preserve Patient #5's organs in an ice-cold preservative solution and have already packed them into sterile containers. These containers contain an icy slush mixture that will help prevent cell deterioration.

Chapter 9
Emergency! Lab 7: Performing Surgery

Recall Questions

1. Why are dissections useful?

2. Why do you think rats were selected as Patient #5, opposed to frogs?

3. What did you find challenging during this lab?

4. What did you like most about this lab?

Critical Thinking Question

1. Although there have been several advancements in technology and medicine, the demand for organs far surpasses the number of organ donors. In order to identify oneself as an organ donor in the United States, a person needs to "opt-in," meaning he or she is not considered an organ donor until he or she takes concrete action to be one. In several European countries, however, a person is considered an organ donor until he or she "opts out" and illustrates an unwillingness to donate. Given what you have learned about the body systems and how easily homeostasis can be disrupted, suppose the United States considered an "opt out" program, would you vote for or against it? Justify your position.

Conclusion
Time to get back to your four original patients!

Chapter 10
The Ominous Phone Call

Filling Out a Prescription

After several days of testing, you have finally developed a diagnosis for each of your four patients. Before releasing them from the hospital you *must* complete the medical charts (Tables 2.1, 2.2, 2.3, and 2.4 on pages 20–23 in Chapter 2) and provide the pharmacist with the following pieces of information under the sections "Diagnosis," "Treatments," and "Prognosis":

1. Results from each of the six tests and a final **diagnosis** of your patient's condition.

2. **Treatments** available for your patient's condition, and

3. A **prognosis**. A prognosis is a medical report dictating a physician's view on a case. It often denotes the chance of a patient's recovery and the doctor's prediction of how that patient will progress. For instance, one might note if the condition is short-term, long-term, fatal, and so on.

Chapter 11
Evaluating the Docs

The head doctor (your teacher) would like your team to present your findings for evaluation and to develop a presentation that encompasses the following information. Be sure to include appropriate data tables and graphs to explain their conditions. When developing your PowerPoint or Public Service Announcement, be sure to collaborate with your fellow doctors. Programs you might want to consider are:

- Google Docs
- Wikispaces
- Prezi

Part I: Oral/Visual Assessment Options

Option 1: PowerPoint (for all four patient cases)

Slide 1: Introduction

a) Introduce yourselves

b) The goal of the project

c) An overview of the four diseases

Slide 2: Lab 1—Urinalysis

a) What was the purpose of this lab?

b) What system of the body is it associated with?

c) How was it done?

d) What could results indicate?

Chapter 11
Evaluating the Docs

Slide 3: Lab 2—Digestive By-Products and BMI Analysis

a) What was the purpose of this lab?

b) What system of the body is it associated with?

c) How was it done?

d) What could results indicate?

Slide 4: Lab 3—Blood Smears

a) What was the purpose of this lab?

b) What system of the body is it associated with?

c) How was it done?

d) What could results indicate?

Slide 5: Lab 4—HIV Test

a) What was the purpose of this lab?

b) What system of the body is it associated with?

c) How was it done?

d) What could results indicate?

Slide 6: Lab 5—Lung Capacity

a) What was the purpose of this lab?

b) What system of the body is it associated with?

c) How was it done?

d) What could results indicate?

Slide 7: Lab 6—Hormone Test

a) What was the purpose of this lab?

b) What system of the body is it associated with?

c) How was it done?

d) What could results indicate?

Slide 8: Patient #1 Diagnosis

a) Describe her medical records symptoms.

b) What was your original hypothesis and why?

c) Diagnosis: What health condition do you think she has now and why?
   Use prescription pad for help.

Slide 9: Patient #1 Health Condition

a) What could have caused her condition?

b) What are the symptoms of this health condition?

c) What are potential treatments?

d) What is the prognosis?
   Use prescription pad and medical school notes for help.

Slide 10: Patient #2 Diagnosis

a) Describe her medical records symptoms.

b) What was your original hypothesis and why?

c) Diagnosis: What health condition do you think she has now and why?
   Use prescription pad for help.

Slide 11: Patient #2 Health Condition

a) What could have caused her condition?

b) What are the symptoms of this health condition?

c) What are potential treatments?

d) What is the prognosis?
   Use prescription pad and medical school notes for help.

Slide 12: Patient #3 Diagnosis

a) Describe his medical records symptoms.

b) What was your original hypothesis and why?

c) Diagnosis: What health condition do you think he has now and why?
   Use prescription pad for help.

Slide 13: Patient #3 Health Condition

a) What could have caused his condition?

b) What are the symptoms of this health condition?

c) What are potential treatments?

d) What is the prognosis?
   Use prescription pad and medical school notes for help.

Slide 14: Patient #4 Diagnosis

a) Describe his medical records symptoms.

b) What was your original hypothesis and why?

c) Diagnosis: What health condition do you think he has now and why?
   Use prescription pad for help

Slide 15: Patient #4 Health Condition

a) What could have caused his condition?

b) What are the symptoms of this health condition?

c) What are potential treatments?

d) What is the prognosis?
   Use prescription pad and medical school notes for help.

Slide 16: Conclusion
Each doctor should have a closing statement that entails one or more of the following:

a) What have you learned from doing this project?

b) How does the subject matter relate to your life or community at large?

c) What advice might you give others regarding their health?

d) What did you find most interesting?

e) What questions did this project raise for you?

Option 2: Public Service Announcement (PSA)—(for one patient case)

Create a public service announcement for one of your patients' health conditions that addresses the questions listed below. Be sure to include evidence to support your claims.

1. What condition is the PSA for?
2. How does one get this condition?
3. What are some of the earlier symptoms?
4. Who should get tested for this health condition, and at what point?
5. How is one diagnosed with this health condition? For instance, what lab tests will they have to do to verify their condition?
6. What are the later symptoms and why? How is the body affected by this condition?
7. Can this condition be treated, cured, or managed? If so, what treatments, cures, or management tips are available?
8. What is the prognosis for a patient who is treated (versus untreated) for your condition?

Part II: Written Assessment Options

To renew your license, you must submit a written portion summarizing your work. Select *one of the two* questions below and write a 1–2 page paper using evidence from your patients' symptoms, the labs performed, and your medical school knowledge to support your ideas.

Option 1

The human body is made up of several different systems. Each system has a separate function, but as you've learned, many of them work together. Below is a list of systems we have discussed in class. Select *two* systems from the list below. Explain how they interact with one another to help keep a living organism healthy and alive. Be sure to cite examples from this project.

- Excretory System
- Digestive System

Chapter 11
Evaluating the Docs

- Circulatory System
- Immune System
- Respiratory System
- Reproductive System
- Endocrine System
- Nervous System *(Note: The nervous system was not covered in this project)*
- Musculoskeletal System *(Note: The musculoskeletal system was not covered in this project)*

Option 2

One's genes and/or lifestyle choices can disrupt homeostasis. Give concrete examples for each of the *two* factors listed above (genes and lifestyle choices), and explain how they may interfere with the health of a human being. Be sure to cite specific examples from this project.

a) Who was affected?

b) What condition did he/she suffer from?

c) How did this condition come about in him/her?

d) What symptoms did he or she experience?

e) What lab tests helped diagnose this condition? Explain how the results were determined.

f) Are any treatments or cures available? If so, what can the patient do to monitor or control his or her condition?

Mystery Diagnosis Rubric for Evaluating the Docs

Doctors are constantly being evaluated on their performance. Great ones may renew their license and are often recommended by patients through referrals, while poor-performing doctors, may lose their license to practice. Today, the head doctor (your teacher) is going to evaluate your performance as a doctor. He or she will determine if you are performing competently or whether or not you may need to be re-evaluated. Below is the rubric for your evaluation.

Using Evidence: The "doctor" can explain the purpose and the methods of the tests he or she performed to make a diagnosis. He/she incorporates all the data to draw valid conclusions.

Student

_____ : <--->

_____ : <--->

_____ : <--->

_____ : <--->

Novice	Competent	Expert
The "doctor" cannot explain the lab in a way that the evaluator can understand the purpose, procedures, nor the results.	The "doctor" appears more confident about the purpose, procedure, and results of some labs over others.	The evaluator has a *clear understanding* of the lab's purpose, procedure, and results based on the "doctor's" explanation.

Chapter 11
Evaluating the Docs

Making Connections: The "doctor" is able to make connections between the knowledge of the body systems, research on the health conditions, and evidence gathered from each lab.

Student

_____ : <-->

_____ : <-->

_____ : <-->

_____ : <-->

Novice	Competent	Expert
The "doctor" does not have a thorough understanding of the body systems and/or diseases; information is inaccurate and/or has difficulty answering questions presented by evaluator.	The "doctor" has a decent understanding of each body system and condition, and sometimes struggled with the questions presented by the evaluator. Most of the information is accurate.	The "doctor" has a *thorough understanding* of each body system and condition; information is accurate, and therefore it is *easy* to address questions posed by the evaluator.

Seeking Significance: The "doctor" understands the importance of these topics and how it affects his or herself, people within his or her community, and those worldwide. He or she can provide recommendations on how to prevent, treat, or manage the conditions diagnosed.

Student

_____ : <-->

_____ : <-->

_____ : <-->

_____ : <-->

Novice	Competent	Expert
The "doctor" does not clearly indicate ways in which this topic is relevant. Lacks suggestions for others on how to prevent, treat, or manage the condition.	The "doctor" is limited in his/her understanding of how this topic is relevant and is limited in his/her suggestions on how to prevent, treat, or manage the condition.	The "doctor" clearly indicates ways in which this topic is relevant and makes suggestions for others to prevent, treat, or manage the disease or condition.

Chapter 11
Evaluating the Docs

Presentation Skills: The "doctor's" presentation style is ...

Student

_____ : <--->

_____ : <--->

_____ : <--->

_____ : <--->

Novice	Competent	Expert
The "doctor" does not act professionally when he/she speaks. The presentation is not revised and the "doctor" seems unprepared to present. Additionally, the "doctor" lacks eye contact and mispronounces many key terms.	The "doctor" could make improvements in one or two areas of presentation; however, these errors do not distract from the presentation itself.	The "doctor" acts professionally when he/she speaks. The presentation contains little to no mistakes. The "doctor" is prepared to present and has great eye contact. He/she has clearly practiced pronouncing key terms.

Written Evaluation: The "doctor's" written assessment is ...

Student

_____ : <--->

_____ : <--->

_____ : <--->

_____ : <--->

Novice	Competent	Expert
The "doctor" does not complete the assignment and/or explanations are unclear. There are major flaws in concept mastery and incorrect use of scientific terminology.	The "doctor" completes the assignment but explanations may be slightly ambiguous or unclear. May contain some incompleteness or a cloudy understanding.	The "doctor" shows clarity of thought and assignment fufills all requirements. The "doctor" shows thorough understanding of scientific content. Statements are supported by evidence.

Chapter 11
Evaluating the Docs

Final Evaluation: The evaluator will determine if the "doctor" can renew his or her license (which indicates a passing grade) or have his/her license revoked (which indicates more practice in this subject area is needed).

Student

_____: Renew or Revoke

_____: Renew or Revoke

_____: Renew or Revoke

_____: Renew or Revoke

Thank you for using this book!

Your opinion is very important.

Please visit
www.StylishSchooling.com
and complete a five-minute
<u>student</u> survey[1] for *Diagnosis for
Classroom Success: Making Anatomy
& Physiology Come Alive.*

---

1. To complete the survey, please make sure you have set up an account. Once logged in, the student survey can be found under "Contact Us → Student Survey." Remember to select "Mystery Diagnoses" from our dropdown menu.

Index

*Page numbers printed in **boldface** type refer to tables or figures.*

A
Abortion by minor, parental involvement and, 70
Alveoli, 62
Arteries, 45, 75
Asthma, 63

B
Bacterial infections, 43, 54
Birth control, **6**
Blood smears (Lab 3), 1, 43–47
    conclusion of, 47
    critical thinking question related to, 46
    lab roles for team members, 44
    materials for, 44
    medical school notes regarding red and white blood cells, **43**
    PowerPoint slide on, 82
    procedure for, 44
    recall questions related to, 45
    recording results for each patient, **45,** 47
Body mass index (BMI), 38
    chart for adults, **37**
    ranges for, **37,** 38
    recording value for each patient, **38**
    relation to health status, 39
    significance of high BMI, 39
Body systems, 85–86

C
Carbohydrates, simple and complex, 38
Circulatory system, 45, 75–76, 86
Code of ethics, 8
Condom use, 49

D
Diabetes, 2, 3
    glucose tolerance test for, 65–66, 68, **68, 68–69, 69**
    information collected on causes, symptoms, and treatments for, **7**
    lung capacity and breathing difficulty in, 58, 63
    type 1 and type 2, **7,** 69
Diagnosis, 1, 2, 11
    evaluating doctors' performance, 2, 81–90
    final, 79
    mystery diagnosis rubric, 87–90
    PowerPoint slides on, 83–84
Diagnostic tests, 1–2
    blood smears (Lab 3), 43–47
    digestive by-products and body mass index analysis (Lab 2), 33–41
    HIV test (Lab 4), 49–55
    hormone test (Lab 6), 65–70
    lung capacity (Lab 5), 57–64
    urinalysis (Lab 1), 25–31
Dialysis, 30

Index

Digestive by-products and body mass index
analysis (Lab 2), 1, 33–41
  body mass index, 38
    chart for adults, **37**
    ranges for, **37**
    recording value for each patient, **38**
  chemical observations for, 35, 38
  conclusion of, 41
  critical thinking question related to, 40
  lab roles of team members, 34
  materials for, 34
  meaning of nutrient indicator results, **36**
  nutrition content found in patients' digestive by-products, **36**
  physical observations for, 35
  PowerPoint slide on, 82
  procedure for, 35
  recall questions related to, 38–39
  recording results for each patient, 41
  setup for, 35
Digestive system, 38, 73–74, 85
Dissection lab. *See* Organ donation surgery

E
Earning your white coats, 1–10
  Hippocratic oath, 1, 8–9
    discussion questions related to, **10**
  information collected on diabetes, **7**
  information collected on HIV infection, **5**
  information collected on pregnancy, **6**
  information collected on sickle cell anemia, **4**
  medical school attendance and graduation, 1, 2, 8
  study group assignments, 3
  study group at the library, 2
  task overview, 1–2
Endocrine system, 65, 69, 86
Evaluating doctors' performance, 2, 81–90
  filling out a prescription, 79
  mystery diagnosis rubric for, 87–90
  oral/visual assessment options for, 81–85
  resources for, 81
  written assessment options for, 85–86
Excretory system, 30, 74, 85
Exhalation, 62

F
Feedback, 2
Fertilization, **6**
Filling out a prescription, 79
Fungal infections, 43

G
Genetic disorders, 43, 86
Glucose tolerance test, 65–66, 68
  graphing results of each patient's test, **68**
  medical school notes regarding normal and diabetic test results, **68**
  recording results of each patient's test, **69**
Glycemic index (GI), 39
Glycogen, 69

H
Hemoglobin protein on red blood cells, in sickle cell anemia, **43,** 46, 58
Hippocratic oath, 1, 8, 11, 30
  discussion questions related to, **10**
  modernized version of, 8–9
Homeostasis, 69, 77, 86
Hormone test (Lab 6), 2, 65–70
  critical thinking question related to, 70
  lab roles for team members, 67
  materials for, 67
  medical school notes regarding human chorionic gonadotropin and insulin, **65**
  PowerPoint slide on, 82
  procedure for glucose tolerance test, 68
    graphing results of each patient's test, **68**
    medical school notes regarding normal and diabetic test results, **68**
    recording results of each patient's test, **69**
  procedure for pregnancy test, 67
    meaning of hCG indicator results, **67**
    recording results of patients' tests, **68**
  recall questions related to, 69
Human chorionic gonadotropin (hCG), **65,** 69
  meaning of hCG indicator results, **67**
  recording results of patients' pregnancy tests, **68**
Human immunodeficiency virus (HIV) infection, 2, 3, **43**

Index

information collected on causes, symptoms, and treatments for, **5**
lung capacity and breathing difficulty in, 58, 63
preventing transmission of, 54
progression to AIDS, 54
susceptibility to other infections in, 54, 58
Human immunodeficiency virus (HIV) test (Lab 4), 1, 49–55
    conclusion of, 55
    critical thinking question related to, 55
    lab roles of team members, 50
    materials for, 50
    PowerPoint slide on, 82
    procedure for, 51–53
    recall questions related to, 54
    recording results for each patient, **53**, 55
    results from limited sexual partner demonstration, **52**
    results from multiple sexual partner demonstration, **52**
    time after initial infection before HIV antibodies show up on, 54
Hypothesis generation, 1, 11

I
Immune system, 54, 86
Infectious diseases, 43, 49, 54
Inhalation, 62
Insulin, **7, 65,** 66, 69

K
Kidney dialysis, 30
Kidney function, 30, 74

L
Lifestyle choices, 86
Lung capacity (Lab 5), 1, 57–64
    acceptable value for height, weight, and gender, 61, **61**
        recording each patient's actual lung capacity compared to, **62,** 64
    calculations for determination of, 59, **60**
    conclusion of, 64
    critical thinking question related to, 63
    determining by a balloon's diameter, **60**
    factors that may cause increase or decrease in, **57,** 58
    lab roles of team members, 59
    materials for, 59
    PowerPoint slide on, 82
    procedure for, 59–61
    recall questions related to, 62–63

M
Medical records, analysis of, 11–19
    for patient #1, 12–13, **20**
    for patient #2, 14–15, **21**
    for patient #3, 16–17, **22**
    for patient #4, 18–19, **23**
    task overview for, 11
Medical school attendance and graduation, 1, 2, 8
Medicare, 30
Musculoskeletal system, 86
Mystery diagnosis rubric, 87–90

N
Nervous system, 86
Nutritional guidelines for school meals, 40

O
Obesity and overweight, **37,** 38, **38,** 40
Oral/visual assessment options, 81–85
    PowerPoint presentation, 81–84
    public service announcement, 85
Organ donation surgery (Lab 7), 2, 71–77
    conclusion of, 77
    critical thinking question related to, 77
    lab roles for team members, 72
    materials for, 72
    procedure for, 72–76
        organ preservation, 76
        preparing for rat dissection, **73**
    recall questions related to, 77
Organ donor programs, 77
Outdoor air quality, 63

P
Pathogens, 43
PowerPoint presentation, 81–84
Pregnancy, 2, 3

Index

information collected on causes, symptoms, and treatments related to, **6**
lung capacity and breathing difficulty in, 58, 62
Pregnancy test, 67
   meaning of hCG indicator results, **67**
   recording results for each patient, **68**
Privacy of patients, 9, 11
Prognosis, 2, 79
Public service announcement (PSA), 85

R
Red blood cells (RBCs), 43, **43**, 45, **45**
   in sickle cell anemia, **43**, 46, 58
Reproductive system, 69, 86
Research task, 1–2
   evaluating doctors' performance on, 2, 81–90
   study group assignments for, 3, **4–7**
Residency programs, 11
Respiratory system, 62, 86
Rubric for evaluating doctors' performance, 87–90

S
Safe sex, 49
School meals, nutritional guidelines for, 40
Sexually transmitted diseases (STDs), 49, 55
Sickle cell anemia, 2, 3, **43**
   hemoglobin protein on red blood cells in, **43**, 46, 58
   information collected on causes, symptoms, and treatments for, **4**
   lung capacity and breathing difficulty in, 58, 62
Sickle cell trait, 46
   sports participation at high altitude and, 46
Spirometer, 57
Study groups in library, 2
   assignment of, 3
Surgery. *See* Organ donation surgery
Systems of the body, 85–86

T
Treatments, 79
   for diabetes, **7**
   for HIV infection, **5**
   in pregnancy, **6**
   for sickle cell anemia, **4**

U
Urinalysis (Lab 1), 1, 25–31
   chemical observations for, 28
   conclusion of, 31
   critical thinking question related to, 30
   lab roles of team members, 26
   materials for, 27
   meaning of nutrient indicator test results, **29**
   medical school notes regarding results of, **26**
   physical observations for, 27
   PowerPoint slide on, 81
   procedure for, 27–28
   recall questions related to, 30
   recording results for each patient, **29**, 31
   setup for, 27

V
Vaccines, 54
Veins, 45, 75
Viral infections, 43, 54

W
White blood cells (WBCs), 43, **43**, 49, 54
   in HIV infection, **43**, 49
Written assessment options, 85–86
   body systems, 85–86
   how genes and lifestyle choices affect homeostasis, 86